MECHANICS' INSTITUTE
MECHANICS'
MERCANTILE LIBRARY

IEE History of Technology Series 5

Series Editor: Brian Bowers

An early history of electricity supply

Previous volumes in this series

Volume 1	Measuring instruments — Tools of knowledge P.H. Sydenham
Volume 2	Early radio wave detectors V.J. Phillips
Volume 3	A History of electric light and power B. Bowers
Volume 4	The History of electric wires and cables R.M. Black

An early history of electricity supply

The Story of the Electric Light in Victorian Leeds

J.D. Poulter

Peter Peregrinus Ltd on behalf of the Institution of Electrical Engineers

Published by: Peter Peregrinus Ltd., London, United Kingdom

While the author and the publishers believe that the information and guidance given in this work is correct, all parties must rely upon their own skill and judgment when making use of it. Neither the author nor the publishers assume any liability to anyone for any loss or damage caused by any error or omission in the work, whether such error or omission is the result of negligence or any other cause. Any and all such liability is disclaimed.

British Library Cataloguing in Publication Data

Poulter, J.D.
An early history of electricity supply: the story of the electric light in Victorian Leeds.—(IEE history of technology series; v. 5)
1. Electric utilities—England—Leeds (West Yorkshire)—History—19th century
I. Title II. Series
338.4'762131'0942819 HD 9685.G73L/

ISBN 0-86341-060-X

Printed in England by Short Run Press Ltd., Exeter

To the memory of my father

Contents

Acknowledgements

The publishers and author are grateful to the following for permission to reproduce the illustrations.
Director of Library Services, Leeds City Libraries: Photographs 1, 2, 5, 6, 7, 8, 9, 10, 11, 12, 13, 14, 15, 16, 17, 19, 20, 22, 23, 24, 25, 26, 32, 33 and 34.
The Locomotive Publishing Co. Ltd: Photograph 3.
E. F. Clark of K.S.L. Publications: Photograph 4.
Hainsworth and Nattress (Photographic) Associates: Photographs 18, 21 and 29.
Yorkshire Electricity Board: Photographs 27, 28, 35, 36 and 37.

Preface

When I first became interested in the history of electricity supply I had no intention of writing a book about it. If I had any intentions at all at that time they were merely to have a few typed pages copied and given to friends and colleagues, but one thing, as they say, seemed to lead irresistibly to another.

It all started when I found some scrapbooks of newspaper and magazine cuttings in a tea-chest in the offices of the Yorkshire Electricity Board in Leeds, where I work. My excitement can be imagined when I found that the earliest scrapbook dated from 10 May 1893 and described the opening of a new Electricity Generating Station in Leeds for the Yorkshire House-to-House Electricity Company. And I soon found that my excitement had been justified, for the books revealed a very strange story – of amazing efficiency by a private company and amusing incompentence by the City Council.

What the scrapbooks also revealed, however, was that they told only a small part of the story. In particular a paragraph in the magazine *Electrical Review* of 19 May 1893 made it quite clear that events had not run smoothly for the City Council in the years prior to 1893:

> 'It must be with a troubled mind that one approaches the history of electric lighting in Leeds. It is a record which, for sheer vacillation and sheer incompetency, is the worst in the whole of the written pages of electricity. It would be idle to attempt to defend the attitude of the Town Council, no excuse can palliate their feebleness of action. They decided upon a definite policy, the first breath of opposition drove them from their position and finally they were constrained to burn what they had adored and to adore what they had burned.'

It was that last phrase – 'to burn what they had adored and to adore what they had burned' – which kept returning and rattling round by mind. Finally it persuaded me that I should seek out the whole story of electricity supply in Leeds. So began a search through the newspapers, magazines and books of the nineteenth century.

I forget now how many years the search lasted but its quite probable that if I had known at the beginning how long it was to last I would never have started! The fact that I perservered (with growing enjoyment, I must admit) is largely due to the help and encouragement I received from so many people along the way.

My grateful thanks are due to all these people but above all I wish to acknowledge the support, both moral and financial, offered by my employers, the Yorkshire Electricity Board, particularly to the West Yorkshire Area Manager, Mr. G. D. Atkinson, and the Leeds Area Engineer (now retired), Mr. R. Jessop. They freely made available the considerable amount of historical material which eventually proved to be in the possession of Yorkshire Electricity Board, they greatly encouraged me in my research, and they made available to me the typing and copying resources of the Yorkshire Electricity Board. I offer thanks here to Dorothy, who was so efficient with both typewriter and wordprocessor.

I sent the first draft to my typescript to the Thoresby Society (the Leeds historical society) and received a humbling, albeit necessary, lesson from Mrs. Rosemary Stevens (Joint Honorary Editor) on how to convert my random jottings into an interesting, yet academically satisfying history. Her criticisms were brutally honest but were also gently encouraging and there is no doubt that it was through Mrs. Stevens that I was able to produce a second draft of the work which was eventually acceptable for publication.

During the re-drafting many people, some of them complete strangers, showed an enthusiasm towards my project which I found delightfully overwhelming. In particular I would like to mention and to thank (in no special order): Mr. Cliff Wimpenny (the Ferranti Archivist) and his Assistant who made me so welcome on my visit there; Mr. E. F. Clark, the owner of copyright material concerning Parsons, who was so courteously helpful and encouraging; Mr. David S. Thornton, Principal Librarian, Central Bibliographic Services at Leeds City Libraries, who allowed me to cause considerable disturbance in the Library while photographing the interior, and who also gave me permission to copy archive photographs and illustrations; Mr. Barry Nattress, who not only took such excellent photographs (with his partner Mr. Hainsworth) but who also gave me important advice on what was possible and suitable; and Mr. J. R. Blakeborough, who helped me to understand the early history of transport in Leeds, and whose interest was vital to my morale.

Chapter 1

The Coloured Cloth Hall

The most popular and best loved politician of Victorian times was undoubtedly William Ewart Gladstone, who had a political career of great brilliance spanning more than fifty years. He was eighty-six years of age when he retired from politics in 1895, and even now, almost a century later, his name and his achievements are well known and well remembered.

There is, however, one incident in his long and excellent career which is in all probability known only to a few: he was – for a short time and completely against his will – elected as a Member of Parliament for the Borough of Leeds. This is an episode well worth recording here, for it was also the influence which resulted in the first vivid impressive public displays of the electric light in the township of Leeds. The strange political involvement of the elderly Liberal statesman in the affairs of Leeds also marked the beginning of the electrical history of the town.[1]

Gladstone: MP for two constituencies

The events leading up to those first displays were unusual and began in 1874 when the Leeds Liberals surprisingly lost a Parliamentary seat in the general election of that year. Their discomfiture was intensified two years later when they again held only one seat in the elections of 1876.[2]

At that time the Liberal and Conservative parties were well balanced nationally but this was definitely not the case in Leeds, which was a veritable stronghold of Liberalism. This local strength was destined to last for a long time and provoked the *Yorkshire Post* to complain, in 1890, that 'not during the present generation has a Conservative ever been deemed good enough to fill the Civic Chair, or even hold a seat on the Aldermanic Bench.'[3] It was not until 1895 that the Conservatives could win enough seats to control the Leeds Council 'for the first time in the history of the reformed municipality.'[4]

With local support thus assured the Liberals confidently expected to hold two of the three seats allocated to the Borough, on a somewhat permanent basis, but it was various apparently irreconcilable splits within the local party which had resulted in their disastrous loss of the Parliamentary majority.

The seemingly fruitless search to find a candidate who could heal the various party rifts ended in 1878 when Gladstone expressed a wish of retiring from his Greenwich constituency. He had decided that at nearly seventy years of age he was no longer able to devote adequate time and energy to both national and constituency affairs and was looking for a more understanding and less demanding constituency. To the Leeds Liberals this was too good an opportunity to miss, so despite Gladstone's protestations that he had no immediate wish to make a decision, he was adopted as candidate for Leeds for the next general election. Gladstone, however, had other ideas and eventually accepted the nomination of Mid-Lothian. His opponent here was to be the Earl of Dalkieth, son of the Duke of Buccleuch, and a hard battle was in prospect, especially as the Duke reportedly controlled a large number of 'faggot' voters. The Leeds Liberals believed that the strength of the Duke would in fact be too great for Gladstone and therefore decided to maintain his Leeds nomination.

Thus, in the election of 1880, the name of Gladstone was a rallying cry – despite his absence – and Gladstone and Barran (the other Liberal Candidate) were elected with huge majorities, both polling double the number of votes of the Conservatives. Gladstone therefore found himself an MP for Leeds, but it was only to be for a short time. At that time polling did not take place simultaneously throughout the land and so it was that the polling at Mid-Lothian took place a few days later and Gladstone was elected there also. He thus found himself in the unusual position of being elected a Member of Parliament for two constituencies at the same time!

In April 1880, after his election at Mid-Lothian, Gladstone wrote to the Leeds Liberals, through their President, James Kitson Junior thanking them for the honour done him but regretting he could not take his seat for Leeds – 'I must regret that my name could not be linked in the annals of your town,' he wrote.[5] He suggested instead that his son Herbert, who had just lost the contest at Middlesex, should take his place, and Herbert did indeed become Leeds MP, being returned unopposed in May 1880.

In August the following year the electors of Leeds were pleasantly surprised to read in the morning papers that Herbert Gladstone had been appointed a Junior Lord of the Treasury, which although an unpaid post was nevertheless considered to be an office of profit under the Crown, a post therefore requiring resignation and re-election. The Conservatives did not put forward a candidate, agreeing with the *Leeds Mercury* that 'the circumstances under which the seat has been rendered vacant . . . are considered sufficient to prevent a contest.'[6]

However, the Liberals decided to take no risks and mounted a full scale election campaign. They advised Herbert to meet his constituents. Accordingly, on Saturday 20 August a large Liberal rally was held in the unlikely premises of the new Crown Point Printing Works, Hunslet, 'a spacious building recently erected by Mr. Alf. Cooke . . . but in which the machinery has not yet been placed'.[7] This was described as 'one of the largest and handsomest public halls in Leeds'[8] and in the event the size was important, for despite the short notice nearly 8,000 Liberal electors were present at the meeting in the large central hall, starting at 4 o'clock in the afternoon. The meeting was obviously expected to be lengthy and it was reported that arrangements had been made 'for using the electric light in case the meeting should not be brought to a close before sunset'.[9]

Fig. 1. *Herbert Gladstone addressing a Liberal meeting in the Cloth Hall Yard, 1 May 1880. This yard was later covered over and turned into the 'Gladstone Hall' for the visit of William Gladstone in 1881 (Leeds City Libraries)*

Strangely, in view of the Press's normal interest in all things relating to the modern novelty of electricity, there are no details of the electric light used at the meeting, but it was probably an arc lamp from the German firm Siemens, for such a lamp had been used in Alf. Cooke's only four days previously to illuminate the evening gala and floral promenade of the Hunslet Floral and Horse Show.[10] The lack of details in this case is particularly strange, for these two appearances of the Siemens lamp – on 16 and 20 August, 1881 – were almost certainly the first occasion on which the electric light was on public display in the town. It was an historic occasion which was largely ignored by both press and public alike.

Historic visit

There was considerable excitement, however, when it was announced during the Liberal meeting that Mr. William Gladstone himself would be visiting Leeds for a weekend in early October. He had long promised to thank the electors personally for the honour they had done him and now at last the visit had been arranged; a weekend was in prospect when Leeds would be 'the scene of one of the most important and imposing political demonstrations ever witnessed in this country'.[11] More importantly the weekend would see the town's first impressive and brilliant demonstrations of the electric light.

The Organising Committee knew that they had a daunting task in making arrangements for the visit for they were aware that the whole country would be inspecting the fruits of their labours. The visit of a politician of Gladstone's stature would evoke an eager response not only from the Yorkshire public but from the nation's Press.

The main problem for the committee was to find and to make ready a hall capable of accommodating the huge numbers of people already clamouring to attend the banquet on the first day and the public meeting on the second. They chose eventually the almost derelict Coloured Cloth Hall, the yard of which was a massive 336 feet long by 100 feet wide, the size of a football pitch. The Hall and yard had been an eyesore for some time and it was later demolished to become the site of the General Post Office in City Square. This was the place chosen; with Victorian nonchalance, confidence and energy, the open yard was converted into a vast covered hall by the erection of a Gothic vaulted roof, made of wood, felt and glass. For a fortnight before the visit, the hall was lighted by temporary 'Brush' arc-lamps to enable the construction and decoration work to go on both day and night. So swift was the transformation from dereliction to beauty, the new 'Gladstone Hall' seemed to have 'sprung into existence as at the touch of an enchanter's wand.'[12]

The town soon filled with people from all over the north of England, many coming with no hope of attending the meetings, and excitement intensified as the date of Gladstone's arrival drew near. By Thursday, 6 October, this infectious enthusiasm had spread to the band of prominent Leeds Liberals who waited to greet Gladstone at the railway station. 'An attempt was made by the members of the Liberal Executive to form into line', reported the *Leeds Mercury*, 'but the endeavour failed owing to their eager desire to pay their regards to the Prime Minister. The line was partially formed, but was so quickly broken.'[13]

From the station Gladstone drove through crowded streets to the home of his host – James Kitson Junior – and the following day was spent at the Town Hall where he lunched and received members of local Liberal constituencies from all over Yorkshire.[14] In the evening, the whole town gathered for the banquet, the fortunate few inside the Gladstone Hall, the rest in the adjacent streets.

The old Coloured Cloth Hall now presented a breathtaking sight. Among the rafters were hung flags of all nations with trophies, festoons and other devices; the

Fig. 2. *Entrance to Coloured Cloth Hall, with yard visible behind (Leeds City Libraries)*

walls were hung with crimson trimmed with white and at both ends of the room were large ebony framed mirrors decorated again with crimson and white. Large devices were mounted on the walls, some bearing the names of principal members of the Liberal Party, one saying 'Free Trade', another announcing 'He revered his conscience as a king', yet another having the message 'Love, honour, obedience, troops of friends'. The tables were covered in flowers, with masses of green foliage and yellow blooms, as were the shelves and recesses. There were palms and other large plants, and in one of the larger recesses the Band of the Grenadier Guards played popular music.[15]

In this amazing Hall dinner was served for 1,500 guests and there was a special gallery for another 400 ladies. So keen was the competition for tickets that many changed hands for two or three times their face value.

Lighting was by 600 jets of Bray's gas burners, mounted in chandeliers suspended from the roof, but the decision had also been taken to use electric lighting of the Crompton/Burgin system, the Brush lights having been removed after their use during the erection of the Hall. The new system consisted of eight 2,000 c.p. lamps supplied from four Burgin machines driven by two steam engines from the Leeds firm of McLarens.[16] This arrangement, of two lamps on each circuit, was so that 'if one machine fails it will have no appreciable effect on the amount of light diffused, and the room, as has occurred at some places, left in darkness.'[17] The Crompton lights were chosen as they were 'the only lamps found to burn with sufficient steadiness to enable telegraph or Post Office sorting work to be carried on effectually, and have been in use for over eighteen months in the Glasgow General Post Office.'[18]

The use of both gas and electric lighting was decided upon partly because of the ability of gas lights to provide heat for the hall and partly because of the work of an influential lobby led by Mr. Woodall (the Leeds Gas Engineer) which did not believe electric lighting to be reliable. They were probably justified too, as it turned out, for it was reported that:

> 'on the first night upon which the electric light was tried there was a sudden collapse in the illumination, to the profound dismay of those who had advocated complete reliance upon it, and who had sneered at Mr. Woodall, our very energetic gas engineer, for having urged that the older but safer mode of lighting should not be entirely neglected. At present it is evident that electricity, though a very brilliant, is not quite a certain light, and that gas must still be employed on great occasions like that of this evening even although it should only be in the humble capacity of hand-maiden to the more novel method of illumination. This I am sure will be consolatory intelligence for the holders of shares in gas companies.'[19]

Despite this apparent failure of the new electric light, the general effect was described as being 'surprisingly brilliant'[20] and the only real disappointment of the evening must have been that Gladstone did not actually arrive at the banquet until the meal was over. Weekends such as this were beginning to be very tiring for the old statesman.

It was unfortunate too that the weather that evening was not kinder, although even a searching downpour could not dampen the enthusiasm of the crowds, waiting patiently to see the torchlight procession.

The procession, when it did set off, must have been an awe-inspiring sight. There were 2,500 torchbearers in the procession and as many were lining the route, while more than 200,000 spectators watched the proceedings with great excitement. The route to Mr. Kitson's residence, Spring Bank in Headingley Lane, took the procession past the house of Mr. Greig – a director of Fowlers, the Leeds steam engine company – at the corner of North Hill Road and Headingley Lane. Here the procession stopped in surprise; the grounds of the house were radiant with electric lights and coloured illuminations, and at the gate was a brilliantly illuminated device with the words:

'Welcome Gladstone
Honour to whom honour is due
Free Trade; Free Labour
Welcome friend of peace.'

The lights had been installed by Hammonds, electrical engineers, and consisted of eleven 2,000 c.p. Brush lights supplied from a dynamo taking $10\frac{1}{2}$ h.p. from a Fowlers compound machine. Globes on the lights absorbed 20 to 25 per cent of the power and moderated the glare of the bare lamps. The illuminations were very much admired and 'the noisy army of satellites raised loud cheers as they passed.'[21]

Other electric lighting of a more public nature was also to be seen in Leeds that weekend. Brush lamps supplied from Brush dynamos had been hung over each entrance door into the 'Gladstone Hall' and in Boar Lane, at its end near the Exchange (the Boar Lane exit from what is now City Square) was erected a great light of 10,000 c.p.[22] This was a particularly interesting installation, the 'Gramme' dynamo being directly coupled to a high-speed steam engine which was especially suitable for this kind of temporary lighting because of its small size and portability and its ability (because of its high speed) to be coupled directly without belts or gears. According to the *Leeds Mercury*[23] this machine had been made and patented by R.C. Parsons but almost certainly the newspaper had got the wrong man – the patentee was undoubtedly C.A. Parsons, who was destined to become one of the famous men in electrical history!

C.A. Parsons

R.C. and C.A. Parsons were two of four sons of the third Earl of Rosse (Richard was three years older than Charles), and at the time of Gladstone's visit both were working for the famous Leeds steam-engine firm of Kitsons. Perhaps the Press can be forgiven their confusion, as Richard was a well respected junior partner in the firm, whereas Charles was almost unknown.

Richard Parsons' main interests outside the firm of Kitsons were connected with pumping engines and the general treatment of water problems; during his partnership he 'interested himself most keenly in the steam tram engines'.[24] Charles, on the other hand, had always had an ambition to produce a high speed engine capable of being connected straight onto a dynamo without the need of clumsy inefficient gears or belts, which could cause slippage, lose power and produce irregularities of the light. After a brilliant mathematical career at Cambridge and an engineering apprenticeship at Sir William Armstrong's works at Newcastle, Charles Parsons joined his brother at Kitsons. He had two main duties while working in the experimental shed: he worked on high speed steam engines and he invented and manufactured torpedoes. This latter was rather a strange duty as Kitsons were not otherwise involved in the manufacture of munitions, and Parsons had not previously evinced much interest in rocket propulsion, but nevertheless both he and his new wife could often be seen in the chill morning mists of Roundhay Park as they tested torpedoes in the lake, very early to prevent interference from sightseers.[25]

It wasn't long though, before C.A. Parsons developed the 'epicycloidal' machine, a high speed steam engine produced for several years by Kitsons and the engine in use in Boar Lane in 1881. The epicycloidal machine consisted of two pairs of

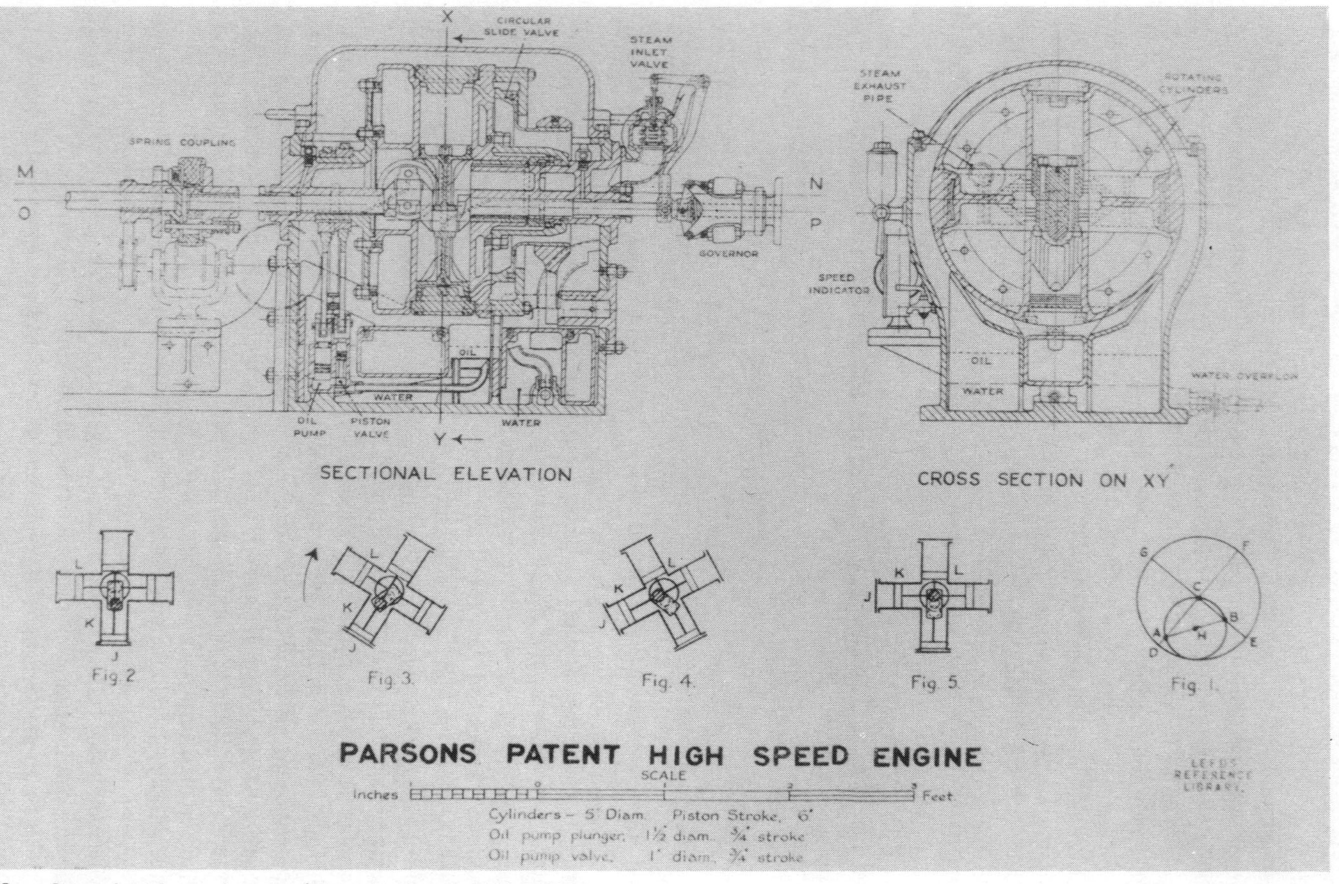

Fig. 3. *Parson's high speed engine* (Kitsons of Leeds 1837–1937)

opposed cylinders, one pair mounted at right angles and adjacent to the other pair in an 'X' shape, the whole mounted in a casing so that all revolved round a pivot central to the X. By revolving, passages into the cylinders were opened and closed to allow steam entry and exit to and from the pistons inside the cylinders, the pistons themselves being centered on a pivot a crank length throw from the cylinder system. The result was that the piston group ran epicycloidally to the cylinder group, making an engine running at 900 rpm without vibration and achieving a compactness which was of a great and novel value. The whole – boiler, engine and a directly coupled dynamo – could be mounted on a four-wheeled chassis, usually about ten or twelve feet in length, which could easily be towed to wherever electricity was required.

The high speed engine was successful for several years as it was useful where space was important, such as on board ship. Several were exported to the Sudan (to be engulfed eventually in sand) and Kitsons used some for their own works. James Kitson Junior was very keen on the electric light, and it was reported that 'the electric light is largely used throughout the works, and is considered to be of special value in the foundry.'[26]

In 1885 Kitsons were awarded a gold medal for the engine at the Inventions Exhibition at South Kensington[27] and other firms attempted to bring out their own design of high-speed engine. Greenwood and Batley, also of Leeds, displayed the high speed Armington–Sims, coupled directly to a Greenwood and Batley 'Jones' 60 lamp dynamo in 1886 at the Liverpool International Exhibition[28] but neither this nor the Parsons machine was to make a lasting impact. Many years later a Kitson executive, E. Kitson Clark, expressed the opinion that although the engines were 'excellent in mechanical contrivance, they were too generous in their appetite of steam, and their very speed, which was undoubtedly a source of valuable education to their author, was not unattended in their earliest form with a certain danger.'[29] Parsons himself agreed with this sentiment, confessing that he 'attributed to Leeds a realisation of the difficulties of high speed reciprocating machinery,'[30] and that the epicycloidal engine did not fulfil his ambition.

After two years at Kitsons, Parsons' wife contracted rheumatic fever (probably from the chilly torpedo mornings), and Parsons resigned from Kitsons in order to take his wife on a long recuperative holiday. On returning, he became a junior partner in the Gateshead firm of Clarke, Chapman and Co. It must have been very much to Kitson's chagrin that almost at once, in 1884, Parsons invented the steam turbine, the ultimate high speed steam engine which was to revolutionise the electricity industry and make C.A. Parsons justifiably rich and famous.

The turbine, running at 12,000 rpm, was the first engine to work at the actual velocity of the steam as it escaped from the boiler[31] and when coupled to a condenser at the steam outlet (as it eventually was) to give a very low steam outlet pressure, the turbine was by far the most efficient steam engine ever invented. Apart from the dynamo, the steam turbine is arguably the most important invention in the history of electricity and has alone made possible the enormous power stations of today.

Excitement at Gladstone meeting

All this fame and fortune for Parsons was well in the future that weekend in 1881 – the hero of the hour was Gladstone, not Parsons. The main event on Saturday, Gladstone's last day in Leeds, was the public meeting in the Gladstone

Fig. 4. *Parson's portable electrical installation (from the collection of the late E. Kitson Clarke, by permission of E.F. Clark)*

Hall, in which Gladstone was to address the public. Overnight the trimmings and decorations for the banquet had been removed and the Hall had been opened up to its full extent. Admission was to be by ticket only, and 25,000 had been sold, but so great was the demand that a healthy blackmarket existed in which sixpenny tickets changed hands for 2/6 or more and 5/– tickets were selling for 10/– or even a sovereign.[32]

The Hall was opened well in advance of the start of the meeting and began to fill at once, and in fact it soon became obvious that the Hall would not be large enough. Many later arrivals, even those with tickets, found that they could not get in. Several Liberal notables were turned away, and went back to the Liberal Club and listened to the speeches on the telephone.

Those who were able to get into the Hall were not always happy with their good fortune, for the overcrowding became so severe that not a few men and women fainted. One stout gentleman was heard to declare, with grim humour, that he had paid 5/– to get in but would willingly pay 10/– to get out. Eventually volunteers had to be called for to climb onto the roof and to tear away large parts of the wood and felt to admit fresh air.

Despite the discomfort, the meeting was a wonderful success, Gladstone's remarkable stamina enabling him not only to survive the ennervating conditions but to deliver a long brilliant speech. Despite his energy, however, Gladstone, now

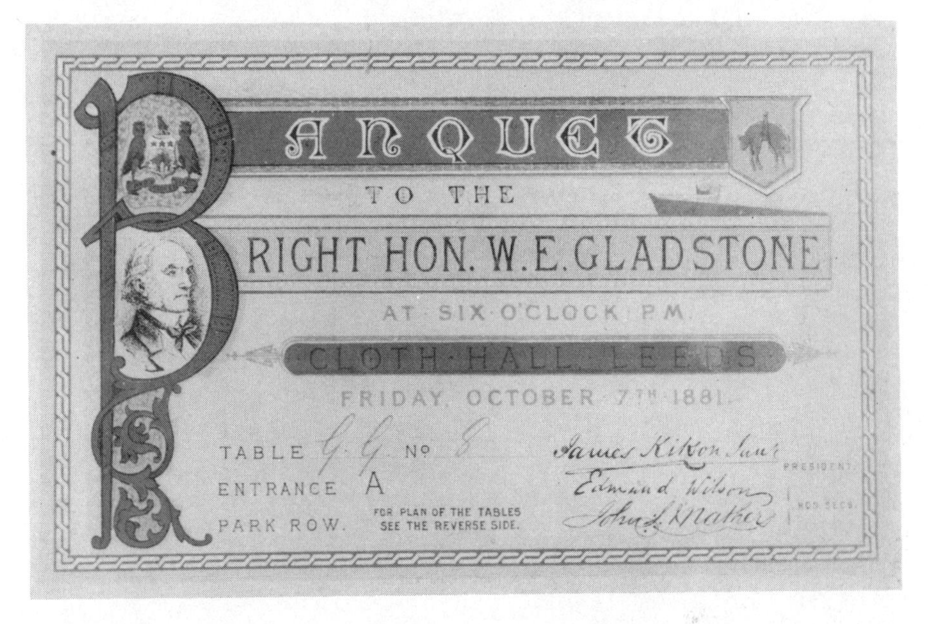

Fig. 5. *Front – ticket to Gladstone Banquet (Leeds City Libraries)*

over seventy years old, must have been thankful at the end of the day to escape to the quiet sanctuary of the Chapel Allerton home of Mr. Barran, one of the Leeds Liberal MPs. Gladstone's public weekend was now over. Dinner at Chapel Allerton was a private affair and the following morning he left Leeds by train.

It had been a weekend of unprecendented enthusiasm and emotion. The Organising Committee had every reason to be pleased with their efforts; it was generally agreed that the townsfolk of Leeds would always look back on the past week with 'pride and satisfaction.'[33] And the electric light? The displays had been more widespread and more brilliant than had ever been seen in Leeds before – there was no doubt the electric light had made a significant contribution to what was 'probably the most imposing demonstration of popular enthusiasm with which even Mr. Gladstone has ever been honoured.'[34]

Notes and references

1. For most of the nineteenth century Leeds was a 'Town' until on 13 February 1893 a Royal Charter was granted by the Queen bestowing on Leeds the status of 'City'.

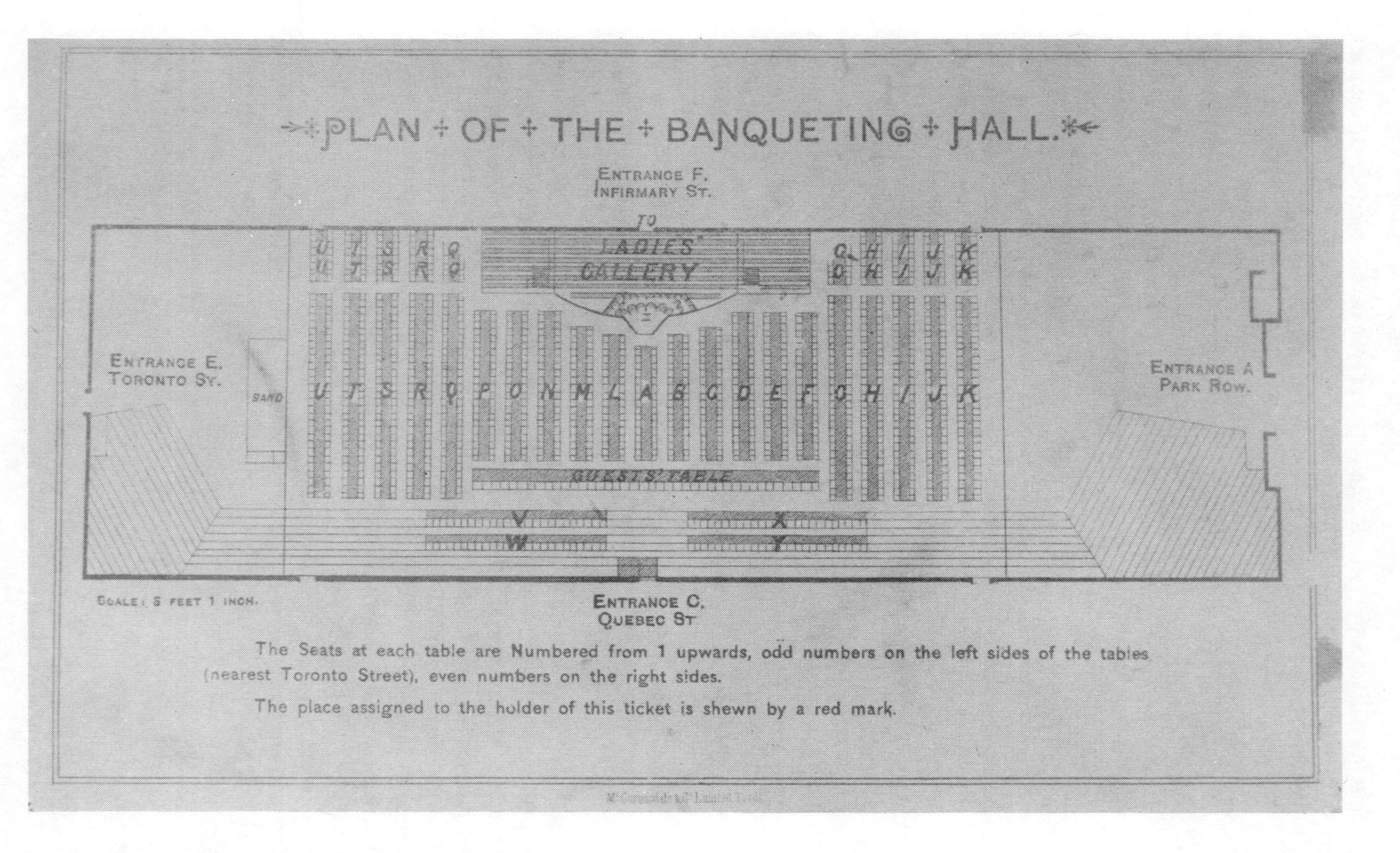

Fig. 6. *Back — ticket to Gladstone Banquet (Leeds City Libraries)*

2. The events leading up to Gladstone's election in Leeds are recounted in the *Leeds Mercury*, 7 October 1881.
3. *Yorkshire Post*, 11 November 1890.
4. *Leeds Mercury*, 2 November 1895.
5. *Ibid*, 7 October 1881.
6. *Ibid*, 17 August 1881.
7. *Ibid*, 20 August 1881.
8. 'Jackdaw' in the Saturday Supplement of the *Leeds Mercury*, 20 August 1881. Apart from its normal Saturday edition the *Leeds Mercury* also published separately a 'Saturday Supplement', a Victorian equivalent to our Sunday colour magazine; although it did contain a summary of the week's news it consisted mainly of puzzles, jokes, household tips, fiction and general interest topics, such as travel tales. Included each week was a column written by an anonymous contributor who called himself Jackdaw. The third verse of Cowper's poem of that name headed the column for many years:

   Fond of the speculative height
   Thither he wings his airy flight
       And thence securely sees
   The bustle and the raree show
   That occupy mankind below
       Secure and at his ease.

   Jackdaw obviously enjoyed the security of his 'speculative height' and he was one of the most strident critics of the Leeds Council, despite the fact that the *Leeds Mercury* was a renowned Liberal newspaper and the Council was for much of the period of Liberal persuasion. For several years Jackdaw had a passionate interest in local affairs but seems to have retired to London, for the column eventually turned from local to national comment.
9. Jackdaw, 20 August 1881.
10. *Leeds Mercury*, 17 August 1881.
11. Jackdaw, 1 October 1881.
12. *Leeds Mercury*, 10 October 1881, an 'Eyewitness'.
13. *Ibid*, 7 October 1881.
14. *Ibid*, 6 October 1881. Ninety addresses were given, mainly from Liberal Clubs and Associations in the north. The first two presented were from Leeds Corporation and the Leeds Liberal Association.
15. *Leeds Mercury*, 8 October 1881.
16. *Ibid*, 6 October 1881.
17. *Ibid*, 8 October 1881.
18. *Ibid*, 10 October 1881.
19. Jackdaw, 8 October 1881.
20. *Leeds Mercury*, 7 October 1881.
21. *Ibid*, 8 October 1881.
22. *Ibid*, 6 October 1881.
23. *Ibid*, 8 October 1881.
24. *Kitsons of Leeds 1837–1937*, E. K. Clark, p. 181.
25. The description of the work of C. A. Parsons at Kitsons, and of the epicycloidal engine is also from the above book, *Kitsons of Leeds 1837–1937*, p. 72.
26. *Engineer*, 7 July 1882.
27. *Yorkshire Post*, 13 August 1885.
28. *Ibid*, 5 August 1886.
29. 'Sir Charles Parsons', by E. K. Clark, in *The Post Victorians,* (1933), p. 510.
30. *Kitsons of Leeds 1837–1937*, p. 74.
31. *Engineer*, 11 December 1885.
32. *Leeds Mercury*, 10 October 1881. Note that all references to money in this story are in the

pounds, shillings and pence (£.s.d.) system, in which there were 12d to a shilling and 20s to a pound. There were thus 240d to a pound. As there are 100p to a pound it can be seen that 1p is equivalent to 2.4d, or 5p equals 1 shilling. A ticket for 2/6d would now sell for 12½p.

33. *Leeds Mercury*, 10 October 1881.
34. *Ibid*, 7 October 1881.

Chapter 2

Technology

With the departure of Gladstone the town of Leeds slowly regained normality; the bunting and the decorations were removed, the electric light installations were dismantled and the Coloured Cloth Hall returned to its normal state of dilapidation. The only electric light to be seen in the town now was in the premises of a few rich or eccentric patrons, such as James Kitson Junior, who used the electric light in his foundry.

There was at that time no publicly available supply of electricity, such as we are accustomed to, for electrical apparatus had not yet reached the stage of development which allowed the easy establishment of public supply networks supplied from central stations. Electricity still had to be generated 'on the spot' where the light was required, so that the few installations which did exist consisted of integral combinations of light, dynamo, steam engine and boiler.

This was a somewhat surprising situation, for electricity is a naturally occurring and frequently observed phenomenon and so had been the subject of intermittent investigation and research for almost two and a half thousand years. The most spectacular manifestation of the presence of electricity is lightning. Each flash is caused by the discharge of accumulated electricity from the clouds to the earth. Another indication of the presence of electricity is the ability of certain objects to attract small particles when rubbed. Modern man-made materials are particularly good at generating electricity in this way, often resulting in an unwelcome buildup of electric charge in everyday articles. This is why vinyl records attract dust and nylon petticoats cling, and is why such stringent precautions are necessary to discharge the electricity safely during the unloading of petrol and oil tankers to prevent a spark from igniting the vapour and causing an explosion.

Electricity caused by friction – by rubbing differing materials together – is called *static electricity* and was first recorded about 600 BC by Greek philosophers, who found that rubbed amber became attractive to light bodies. The Greek name for amber (electron) is the root from which our word electricity is derived.

Having recorded the phenomenon, the Greeks seem to have lost interest in electricity and it was more than two thousand years later that scientists again began the serious investigation of friction generation. By the middle of the eighteenth century generators of static electricity had reached considerable proportions, but there was little public interest, for static electricity has no application of benefit to the public and has no place in electricity supply. The problem with static electricity is that it only exists in the form of a static charge, which when discharged releases its energy almost instantaneously, like the flash of lightning – the energy, alas, cannot be controlled. The electricity which we use for heat, light and power, on the other hand, requires *current electricity*, or electricity which flows regularly and continuously, which can be turned on or off at will and which can be increased or decreased to suit the duty required of it.

Current electricity and the first electric lights

The discovery of current electricity was an accident, as is often the case with important scientific discoveries, and resulted from a series of experiments which took place in Italy at the end of the eighteenth century. These experiments involved the effect of static electricity upon the muscles and nerve fibres of small animals and birds, principally by an Italian surgeon named Beccana (whose experiments on cockerel's legs were recorded by George Adams in 1784) and by Galvani, a Professor of Anatomy, whose frog leg experiments of 1786 were published in 1791.[1] These experiments showed that under certain circumstances dissimilar metals in contact with each other produced small amounts of electricity, evidenced by the contraction of the stimulated muscles. This in turn led a third Italian, Volta, to realise that electricity could be generated chemically and led to his development of the first chemico-electric battery – the Voltaic Pile. This consisted of a pile of alternate discs of two dissimilar metals, such as copper and zinc, or zinc and silver, separated by pieces of flannel or pasteboard moistened with salt water or with water acidulated with sulphuric acid.[2] Although giving no more than a trickle of electricity, it made available a source of current electricity which flowed continuously (until the pile or battery drained) and could be regulated. In commemoration of his achievement Volta's name has been perpetuated in our name for the unit of electrical pressure – the Volt.

The cells which Volta produced supplied only a small amount of power and there was of course no immediate use for electricity; it was merely another tool available to scientists, until, that is, there occurred the second major breakthrough in the story of electricity supply. The *Philosophical Magazine*, describing a lecture at the Royal Institution in 1810, tells us in wonderment that: 'the spark, the light of which was so intense as to resemble that of the sun, struck through some lines of air, and produced a discharge through heated air of nearly three inches in length and of a dazzling splendour.'[3]

The magazine was describing the *electric arc*, discovered by one of England's scientists – Sir Humprey Davy – a man of huge and varied talents about whom the poet Coleridge is supposed to have declared that 'if Davy had not been the first chemist, he would have been the first poet of his age.'[4]

It was actually a Frenchman, Curtet, who had found that breaking a circuit carrying electric current produced a brilliant spark, but Davy, using a huge battery of 150 cells, had then discovered that he could sustain an electric arc between carbon pencils when he slowly separated the tips. To increase the intensity of the arc, Davy vastly increased the size of the batteries until, by courtesy of a fund organised by a few members of the Royal Institution, he was able to build a battery of 2,000 double plates. It was this colossal battery which was used in the public lectures reported by the *Philosophical Magazine*.

The excitement of the *Philosophical Magazine* may seem strange to us now but it was an excitement which was undoubtedly shared by the readers of the magazine. An artificial light of such brilliance was then unknown, indeed artificial light of any sort was rare in the early years of the nineteenth century, so that the scientist Sir Joseph Swan was prompted to call them the 'dark ages'. He was himself a child of those times, being born in 1828, and in that year, he tells us:

> 'the rush light, the tallow dip or the solitary blaze of the hearth were common means of indoor lighting, and an infrequent glass bowl, raised eight or ten feet on a wooden post and containing a cupful of evil-smelling train oil with a crude cotton wool stuck in it, served to make darkness visible out of doors. In the chambers of the great, the wax candle or, exceptionally, a multiplicity of them, relieved the gloom on state occasions, but as a rule the common people, wanting the inducement of indoor brightness . . . went to bed soon after sunset.'[5]

Davy had obviously found the one use for electricity – the *electric light* – which would find favour with everyone, and yet ironically Davy himself seems to have been more interested in the heat of the arc than he was in its light; many materials could now be fused which before had resisted fusion and a whole new range of experiments opened up to him.[6]

To those who were interested in the light, however, it was soon apparent that the electric arc could not readily be converted into the electric light, because there were two particular difficulties which prevented its practical application. First, the carbon points, made of soft charcoal, soon burned away and caused an annoying frequency of carbon replacements. Second, it was difficult to keep the carbons so adjusted that the arc struck and then remained constant. This was a complicated requirement in that when the lamp was not working the points had to be touching so that when the lamp was switched on, by completing the circuit, the current would flow. It was then necessary to separate the points to strike the arc and then to maintain a constant gap between the points, thus maintaining a constant arc. This latter point was very important in an arc lamp for any variation in the arc would cause an irritating variation in the light. The difficulty of carbon regulation was increased if the carbons burned away quickly.

Work to overcome these difficulties was carried on principally in France, although W. E. Staite, in England, obtained various patents for the manufacture of carbons and regulators and achieved considerable local fame. On both sides of the Channel there were attempts to manufacture carbons which were both clean burning and long lasting and various carbon materials were tried, often mixed with substances thought to contribute clean carbon, such as treacle or sugar.

In 1843 Foucault, for example, made use of the carbon from the retorts of gas works[7] and nearly forty years later similar methods of manufacture were being carried out by the French. In 1882 we are told by the *Engineer*[8] that:

'carbons for arc lamps are nearly all made in France. The process is kept a secret as to details by the most eminent makers. Generally speaking, it may be said that carbons are made by grinding very good and clean coke to a fine powder and mixing it into a plastic mass with treacle or syrup. This mass is then made into rolls, each of which is placed in a strong iron mould and subjected to a pressure of many tons by hydraulic presses. The rolls are then packed in seggars or fire-clay boxes, which protect them from the air, and they are then baked at a red heat for several weeks.'

During use, the article added,

'the carbons waste slowly away, the positive fastest. The rate of waste depends on various conditions. In good lamps a 5/8 inch carbon will waste at the rate of 1 inch per hour.'

Early lamps were regulated by hand, but automatic systems soon appeared, some operated by clockwork, some working on the 'pyrometric' system (where alterations to a gear train were stimulated by a wire affected by the heat of the arc) and the more sophisticated operated by a closed-loop electro-magnetic system, in which an operating solenoid was activated by the varying current flowing through the arc. No matter which system was used, however, perfect regulation was never achieved and a flicker, or 'wink', was typical of arc lamps.

Michael Faraday and electricity generation

Despite the relatively advanced development of these lamps, however, half way through the eighteenth century, they never actually achieved the public acclamation that they deserved, for one unfortunate but simple reason – there was no proper supply of electricity available. Voltaic piles (and a few batteries developed from these piles) were still the only source of electricity and were weak and of low capacity. They were highly uneconomic, too, for they used zinc, silver or similar metals, and zinc, was, as Professor Tyndal pointed out 'an exceedingly expensive fuel.'[9]

The principles of electricity generation – to obviate the need for batteries – had been established in 1831 by the experimenter, Michael Faraday, but he, like Davy, was interested more in science than invention and he made no attempt to apply his principles to the practical manufacture of commercial machines. Indeed, the construction of commercial dynamos proved extremely difficult for all and it was almost thirty years later that the first generator/lamp sets were successfully demonstrated.

Davy died in 1829 and so never witnessed the triumphant experiments made by Faraday in the latter half of 1831, experiments which were to ensure everlasting fame for Davy's one-time pupil (Faraday began his scientific life as Davy's assistant at the Royal Institute). The experiments lasted for ten days which were spread over a period beginning in August 1831, culminating on 24 November when Faraday read a memoir to the Royal Society. The results were published in January 1832 and were destined to change the world.

Faraday had long known that an iron bar became magnetised if a copper coil carrying an electric current was placed round the bar. He therefore believed (as indeed did many other scientists) that if electricity caused magnetism, then surely the converse should be true. As early as 1822 he had carried out experiments to try to produce this electricity, but with no success, and over the next few years tried various other experiments, but always without success. His fifth attempt (since 1822) began in the summer of 1831, when he had made a six-inch soft iron ring on which he wound copper coils at opposite sides of the ring. Across one coil he connected a battery of ten pairs of plates, across the other a galvanometer. To his disappointment he found that nothing happened, until he noticed that when he connected or disconnected the battery there was a deflection of the galvanometer. This encouraged him to connect bigger batteries, until with a battery of 100 pairs of plates the galvanometer needle was caused to spin violently for a short time. Even more exciting, when the galvanometer was replaced with a pair of charcoal pencils so placed that there was a small gap between their tips, the switching on and off of the battery caused a small spark to be produced between the points. This ring was, of course, the first transformer.

Faraday then tried various combinations of coils, some on cores, others not, with similar results. The fifth day of experiment, on 17 October, he took a copper coil (with a galvanometer across its ends) into which he inserted, then removed, a magnet. He found that the galvanometer needle moved one way when the magnet was inserted, the other way when it was pulled out.

These experiments persuaded him that electricity could be produced in a coil in two ways. First a changing current in a coil produces a current by induction in a similar coil, especially if both are mounted on the same iron core. Second, current is produced in a copper coil when a magnet is within the coil, provided there is relative movement one to the other.

These discoveries caused him on the ninth day to make a copper disc turn round between the poles of the great horse-shoe magnet of the Royal Society, when a galvanometer connected to the axis and edge of the disc moved as the disc turned. A copper wire drawn between the poles and conductors produced the same effect.

Faraday published details of his experiments (in a different order to their actual execution) as 'Experimental Researches in Electricity' and subsequently constructed many simple machines, although none were large enough to give off useful amounts of electricity. Similar machines, mostly for experimental or demonstrative purposes, were made over the next few years by Pixii, Saxton, Clarke, Wheatstone, Cooke and Woolrich. One machine was capable of decomposing water to provide hydrogen and oxygen, another was used for electroplating and a third (handcranked) was used on the telegraph system, but none were powerful enough to give proper supply to arc lights.[10]

The first generators

It was twenty-two years after Faraday's discoveries that the first generator for the electric light was produced. In 1853 an Englishman, Professor Holmes, took over a machine designed by Abbé Nollet, Professor of Natural Philosophy at the Military School in Brussels, and this was followed in 1859 by a similar machine engineered by Joseph Van Malderon, who had previously assisted Holmes.[11]

These machines were both magneto generators; a coil wound on an armature was rotated in an electric field provided by permanent magnets mounted in a stationary frame surrounding the armature (the field magnets). Having permanent magnets, the fields were weak and the machines were in consequence very large and inefficient. A typical Holmes machine stood nearly four feet tall and weighed a massive two tonnes[12] and yet could supply only one arc lamp, an inefficiency whch in practice restricted the application of these machines to lighthouses, where the brightness of the lamp compensated to some extent for the inefficiencies. The brightness, in fact, was a pleasant surprise to the Elder Bretheren of Trinity House, the arc lamp being some twelve and a half times brighter than the finest Trinity House six wick oil lamp.[13] Such machines were obviously of no real practical public application but it was hard to see how they could be made smaller.

Then in 1866 the solution was presented simultaneously (but independently) by Siemens, Wheatstone and the Varley brothers – the self-excited electro magnetic machine, in which the permanent magnets were replaced by electro-magnets, supplied by the electricity generated by the machine itself, fed through the electro-magnets before being supplied to the load.

The principle of electro-magnetic fields in a machine had actually been first patented by Wheatstone and Coope in 1845[14] but the idea was never taken up, although Wheatstone later used a magneto machine to supply the field of an electro-magnetic machine, comparing and proving the larger output of the second machine. But it was difficult to see how his idea could be used economically until the discovery of the self-excited machine, which cleverly provided a supply to the electro-magnets without the need of provision of another source of electricity.

The operation of the new self-excited machine was very simple. When starting the dynamo there was no current flowing in either the armature or the field coils, but the residual magnetism always present in the iron to some degree produced infinitesimal currents in the armature as it turned. These currents were then carried round the magnet, increasing very slightly its power, and in turn increasing the size of the current produced in the armature. So the currents of both armature and magnet would grow in turn, until the current was of such a size that the magnet had reached a state of saturation and would produce no extra magnetism.

The characteristics of this series-would generator, as it was called, were such that it gave out almost constant current, independent of the external resistance, a characteristic ideal for use with arc lamps for two reasons:

1. There was a continual danger in an arc lamp that the rods would touch, indeed this was necessary to strike the arc. If the current was not held constant, a large overload would occur which could damage the dynamo and connections, with the danger of fire.
2. The arc resistance in an arc lamp is inversely proportional to the current which can lead to unstable operation. If, for example, the arc resistance happens to decrease slightly, the current increases, the resistance decreases more and so on, until a large current is produced giving rise to the same dangers as above.

The Gramme dynamo and Jablochkoff candle

The first self-excited generators – and others made in the late 1860s – were never really satisfactory in their output and it was Z. T. Gramme (a Belgian who had once worked for Societe de l'Alliance) who is credited with producing the first really practical generators.[15] The Gramme Machines were based on the 'ring armature', first invented by the Italian Pacinotti, but Gramme also made various improvements to his machines which eliminated sparking and heating and thus enabled the machines to run with a larger power output.[16] These were notable achievements and prompted Professor Tyndall to comment that the machines were of 'exceeding beauty and exceeding power'.[17] In France, where the electric light was used most extensively, all the machines began to be replaced by Gramme dynamos.

In 1872 the first Gramme dynamo was introduced into England, when a Mr. Werdermann brought one from France for the use of Mr. Conrad William Cooke, a civil consulting engineer.[18] The probable first use of this dynamo was the following year, when Mr. Cooke installed it in the clock tower of the House of Commons, to supply a Serrin arc lamp (also from France) which acted as a signal lamp to show when the House was sitting. Because of this the first Commissioner of Works made it a *sine qua non* that the lamp must not go out, so it was adapted so that the carbons were able to be changed without turning off the light. The installation was not a success, however, due to the expense. This was mainly due to the fact that a skilled man was needed to tend the lamp, and having to spend so much of his time going up and down the tower steps, he had little time for other duties. Within a short time the lamp was removed and replaced with a gas light.

In 1876 a Russian Officer – Paul Jablochkoff – introduced an invention which ideally complemented the steady Gramme dynamo: it was a revolutionary new arc lamp called the Jablochkoff Candle. Unlike ordinary arc lamps, in which the rods were placed tip to tip in a straight line, the rods of the candle were placed side by side, separated by a plate of kaolin, the tips of the rods being connected by a bridge of carbon paste which formed an electrical connection between the two rods so that current would flow when the supply was switched on. The carbon paste then soon burned away leaving a gap between the carbons across which the arc formed. As the carbons wore away the kaolin plate melted so that the arc was always maintained between fresh carbon.

Fig. 7. *Gramme machine 1874 (Appendix to 'Report of Select Committee on Lighting by Electricity, 1874)*

It was essential that the two rods wore away equally, to maintain a constant arc length and this was found to be difficult when DC current was used, as the positive rod wore away much quicker than the negative. To overcome this the positive rods were made approximately twice the thickness of the negative rods, until it was realised that rods wore evenly when AC electricity was used. When used with AC generators, therefore, the candles worked perfectly with equal sized rods and were soon a huge success. The great advantages were the lack of a regulator and the great steadiness of light, and it was soon found, too, that the candles had another great advantage – they had a brightness of only some 500 candles, compared to the tens of thousands of candles of ordinary arc lamps and were therefore much more convenient in use. They took less power, as well, and several candles could be operated from the same machine.[19]

The three years from the end of 1876 to the beginning of 1880 saw a tremendous increase in the popularity and installation of the Jablochkoff–Gramme combination, although the growth was virtually all in Paris, carried out by French

Engineers.[20] There was a reluctance to install electric lighting in England which caused the *Electrician* to bewail the fact that although the use of the electric light was extending in Paris, 'yet in London there is not one such light to be seen'.[21] Towards the end of that year, 1878, the *Engineer* also felt obliged to lament this lack of progress. 'Thousands of people will witness the experiments now being made in Paris, and we may be well assured that many of these will not come home to ill-lighted streets and hold their tongues.'[22] Within a few short months of these comments, however, the electric light was to be seen in several places in England (mainly in London) as Jablochkoff candles were installed in the Gaiety Theatre, Billingsgate Fish Market, Thames Embankment,[23] Holborn Viaduct, various railway stations and factories, and as an experiment in the British Museum (which normally closed as the sun went down).

The Billingsgate installation was not a success, partly because the electric light did not give off the same warmth as the gas had done, and partly because the traders were not used to having their wares so well illuminated![24] The Holborn Viaduct installation, too, was unsuccessful, mainly, it was stated, because the French workmen were continually intoxicated – 'through being treated so often,' as the French Chief Engineer explained. The lights went out altogether on 10 January when the attendant of the steam engine fell asleep and let the fire go out.[25]

The Jablochkoff candle enjoyed a growing success for a few years, but gradually there came the realisation that the very brightness of the arc lamp which had so attracted the public was actually preventing its further application. Lights of 500 or more candles were obviously ideal for illuminating large open spaces, or the interiors of factories, or large public rooms, but they were much too bright to be used in people's homes or in small offices and shops. They could never be suitable for normal indoor use. This presented a major problem for the supporters of the electric light, who realised that such a bar to the most lucrative part of the market would effectively condemn electricity to a minor and unprofitable role in the field of lighting. There seemed to be little prospect of investment or development unless someone could solve the 'subdivision of the light', as it became known, and this became the major task for many of the world's scientists and inventors.

The incandescent lamp: Swan and Edison

There were those, of course, who were relieved to realise the limitations of electricity. A meeting of the London Gaslight Company, for instance, agreed in October 1879 that they could 'afford to laugh at the electric light now',[26] but alas, no sooner had their scorn been recorded than the final major electric light invention – the incandescent lamp – began to be demonstrated around Britain.

The principle of passing a current through a wire of filament to make a glow had been known and demonstrated for many years, but what few people realised was that to convert the glow into a light the filament had to be made with a high resistance, to increase the power, and it had to be placed in an oxygen free atmosphere, preferably a vacuum, to prevent the filament burning away. Two men who did realise this were Swan, in England, and Edison, in America, and both worked without reference to each other, although Swan had a good idea of what Edison was doing, as his work was widely reported on.

Swan was a chemist, a partner in the photographic firm of Mawsons, in Newcastle, so the creation of a suitable filament caused him no great difficulty. Like Edison, though, he could not evacuate the glass bulbs sufficiently well to prevent the filament burning away, vacuum technology then being in its infancy. Eventually however, after a long twenty year wait, he met a Mr. Stearn of Birkenhead, who was an expert amateur in working with vacuums and using the newly developed Sprengel pump he at last succeeded in producing a light bulb that worked, and demonstrated it is public in Newcastle before the Literary and Philosophical Society in February 1879. Later that year both his house and the house of Sir William Crookes were lit by the new lamps. Swan subsequently arranged demonstrations of his lamps all over the country and on 1 March 1881 he gave a lecture and demonstration at the Leeds Philosophical and Literary Society in the Philosophical Hall, Park Row. His lights were demonstrated successfully, and a 6 h.p. 1,500 rpm Kitson engine was used for the first time with Swan lamps, giving a very steady light due to the steadiness of the engine. It is recorded that 'a brilliant white, but by no means dazzling light was emitted. The heat from the lights was almost nil.'[27]

Edison, despite the fact that he was one of the world's most capable inventors, with a laboratory and staff to aid him in his work, was slightly delayed in his project. Although there were constant reports that he had solved the problems, he was not able to demonstrate a successful lamp until October 1880 – nearly twenty months later than Swan.

Despite the pleadings of Mr. Stearn, Swan had never taken out patents to protect his incandescent lamps, as he considered the principles of incandescence merely natural scientific laws and therefore unpatentable. It must have been something of a surprise to him, therefore (but not to the worldly Mr. Stearn) when Edison took out a patent phrased in the broadest of terms – the incandescent lamp possessing as its cardinal features a carbon filament within a glass receiver from which the air had been exhausted (British Patent No. 4576–November 1879).[28]

In 1882 the Edison Company applied for an injunction to prevent what was now the Swan United Electric Lighting Co. Ltd, from making incandescent lamps. An interlocutory injunction was refused and both companies were faced with extensive and protracted legal argument, for Swan had a good claim for 'prior usage', which could have invalidated the Edison patent. Rather than face the risk and expense of doubtful legal action, however, both companies agreed upon an amalgamation and so was formed the Edison and Swan United Electric Lighting Company, in October 1883.[29]

Strangely, Swan himself had to argue against his own claim for 'prior use' when the new company took legal action against various other companies who were, it was claimed, infringing the Edison patent, now in the possession of Edison and

Fig. 8. *Swan incandescent lamp – bulb and connections ('Electric Lighting: Report of Joint Committee, Leeds 1882)*

Swan and providing a lucrative British monopoly in incandescent lighting. In order to overcome the Swan 'prior usage', which if proved would have destroyed both the patent and the monopoly, the company had to prove either that the early Swan lamps (of 1878 and 1879) were unsuccessful experiments, or that carbon filaments had not been used. The dispute eventually reached the Court of Appeal, who ruled that the early Swan lamps had not used filaments – the carbonised paper strips were too big to be called filaments – and the patent was therefore upheld. Poor Swan, although in this action protecting his financial interests, had thereby unjustly diminished his own inventive responsibility. It was Edison instead who took all the credit for the invention of the incandescent lamp.[30]

The introduction of the new 'sub-divided' lights completed the apparatus needed to give universal supplies of the electric light and those rich individuals, businesses and manufacturers who could afford the expense of the new machinery began to install their own electric light systems, on their own premises. Public centralised supply was not yet a practical possibility, however, as a Parliamentary Select Committee[31] discovered in 1879.

This committee had been set up to investigate the possibilities of lighting by electricity, but soon reported its opinion that the immediate use of the electric

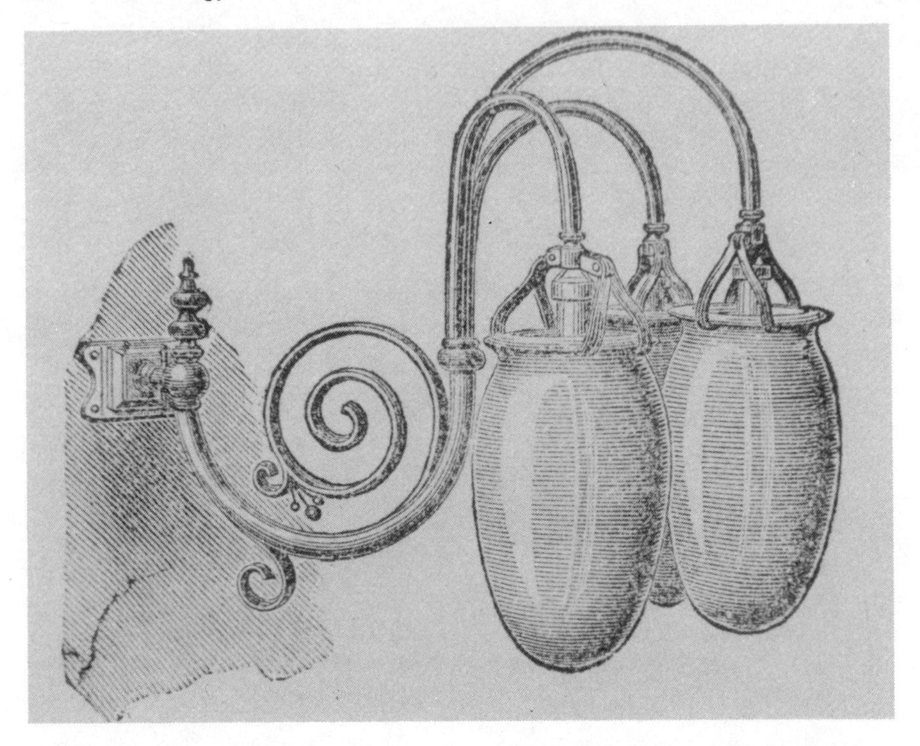

Fig. 9. *Swan incandescent lamp – typical bracket and globes ('Electric Lighting: Report of Joint Committee, Leeds, 1882)*

light was difficult and unlikely. This opinion was formed mainly because of the committee's realisation that electrical apparatus was not yet technically developed to a point where widespread central supply was an economical or practical possibility. But there was also a recognition that a major political obstruction prevented supply to the public – no one had the right to lay cables in public streets to carry electricity for sale! It's true that local authorities had indisputable right to break up public highways to lay pipes, drains and cables, but this was limited to purposes associated with public health, such as sewerage and water supply and they were not allowed to lay apparatus in the ground for profit, unless expressly authorised by Act of Parliament. Local authorities were therefore in the position of being able to lay cables to provide electricity supply for lighting the streets but not for sale to the public.

Municipalists versus capitalists

This restriction obviously prevented economical supply to the public. The committee, in recognising this bar to development, recommended that ample powers should be given for the laying of electric cables, whenever this should prove desirable, but they added – somewhat controversially – that these powers and the subsequent powers to control the distribution and use of electricity should be placed in the hands of local authorities, and only given to private enterprise failing the acceptance of the local authority, and then only for the short period sufficient to remunerate the private company for its enterprise. In other words, the Playfair Committee had committed itself wholeheartedly to municipalism.

Although the nineteenth century had up to that time been a period of encouragement for private enterprise in Britain, a time of *laissez-faire* politics, the Playfair Committee's municipal ideology was representative of a point of view which had been growing in strength throughout the land. There was a rapidly intensifying public dislike for the private companies which ran many of the essential public services, such as water and gas supply and public transport. There were many expressions of dissatisfaction with the poor service and high charges of many of these companies, compounded by a disturbingly high-handed and arrogant attitude. The Playfair Committee were merely reflecting the opinion of a large selection of the community when they recommended that electricity supply should be controlled by municipal authorities, for the benefit of ratepayers, rather than by private companies for the benefit of rich shareholders.

Unfortunately, despite their protestations, many local authorities had not the slightest wish to become involved in this new and potentially disastrous technology. There was no evidence at that time that the electric light would ever become practicable, economical, necessary or popular and there was many a town councillor who baulked at the thought of involving ratepayers' money in such a dubious venture.

From this time on, until nationalisation in 1947, the growth of the electricity supply industry came to be effected more by the arguments between municipalists and capitalists than it did by the development of the apparatus, important as this was. In those countries where there were no political barriers to development, such as the USA, electricity rapidly became popular and the electricity companies became successful, at home and abroad. In Britain, however, arguments and petty jealousies delayed success.

Not that there was disagreement throughout the whole of Britain, for development and growth varied from place to place, depending to a large extent upon the political persuasion of the local authority, its strength of will and the money it had available for investment (or, perhaps, speculation).

Some local authorities were determined to control electricity supply from the start, such as Bradford, which in 1889 became the first municipal authority to open its own electricity supply station. Others, particularly in London, were happy to allow private companies to take the responsibility. Some authorities did nothing, being too small to be of interest to private companies and too poor to be able to afford an undertaking of their own.

Often, though, decisions, were not so simple. Many councils, despite a strong commitment to municipal control, found themselves having to accept a private undertaking within their area, and later were forced to consider an expensive take-over to restore municipal management to what had become a successful public service. It was the experiences of such councils which best illustrates the struggle between municipalism and capitalism, a struggle which so delayed and diminished the development of the electricity supply industry in Britain.

The events which occurred in Victorian Leeds were typical of the experiences of many of these councils. Circumstances were not always identical, of course, but the political and financial problems encountered by Leeds Town Council were similar to the problems encountered in other councils. And the personal foibles and inadequacies of the Leeds Councillors (which influenced events so greatly) were surely matched in the personalities of councillors throughout the land. There can be little doubt that the incidents which contributed to the development of the electric light in Leeds represented development nationally: to understand the story of electricity supply in Leeds is to understand the story of electricity supply in Britain.

Notes and references

1. 'Physiological radio receivers', by V. J. Phillips, *IEE Electronics and Power*, May 1981, p. 409.
2. *The British Encyclopaedia*, (London, 1933), Volume 10, p. 472.
3. *Philosophical Magazine* quoted in 'Electric Arc light', by R. J. Simpson and Prof. H. M. Power, *IEE Electronics and Power,* September 1979.
4. *The British Encyclopaedia*, Volume 3, p. 432.
5. *Sir Joseph Swan*, M. E. Swan and K. R. Swan, (Newcastle 1968).
6. *Philosophical Magazine* quoted in 'Electric Arc Light', *IN: IEE Electronics and Power*.
7. *Ibid*.
8. *Engineer*, 7 April 1882.
9. Report by Select Committee on Lighting by Electricity (Playfair Committee Report), 1879, p. 3.
10. *A. History of Electric Light and Power*, B. Bowers, (UK and New York 1982), pp. 72–77.
11. *The Early Days of The Power Station Industry*, R. H. Parsons, (1940), pp. 1–2.
12. *A History of Electric Light and Power*, p. 77.
13. Playfair Committee Report, p. 55.
14. *A History of Electric Light and Power* p. 84.
15. *Ibid*, p. 90.
16. *Engineering*, 27 November 1874.
17. Playfair Committee Report, p. 8.
18. *Ibid*, p. 35.
19. *Ibid*, p. 87.
20. In 1879 there were 500 lamps in Paris, 250 in the rest of France and 800 in the rest of the world (Playfair Committee Report, p. 89).
21. *Electrician*, 20 July 1878.
22. *Engineer*, 23 August 1878.
23. It was probably this original installation on the Embankment which made such an impact on the Leeds artist, Atkinson Grimshaw. One of his paintings was entitled 'The Electric Light on the Thames Embankment' which was – according to an advertisement in the Leeds Mercury of 17 April 1882 – 'a view of the Thames at London, showing Cleopatra's

Needle; a remarkable night scene in which the artist has aimed with great success at giving the effects of the electric light and gaslight side by side.'

24. Playfair Committee Report, p. 82.
25. *Ibid*, p. 81.
26. *Leed Mercury*, 10 October 1879.
27. *Ibid*, 5 March 1881.
28. *Sir Joseph Swan*, p. 69.
29. Edison insisted on his name being first – Swan accepted the order of the names because of the alphabetical priority.
30. *Sir Joseph Swan*, pp. 101–102.
31. Select Committee on lighting by electricity – ordered on 28 March, sat on 17 April, reported on 13 June 1879. Chairman was Dr. Lyon Playfair, hence the name of the committee, the 'Playfair Committee.'

Chapter 3

Yorkshire Brush Electric Light and Power Company

The usual source of the electric light in 1881 – at the time of Gladstone's visit – was the arc lamp, which despite its 71 year development period (since the demonstration in 1810 by Davy) was not yet satisfactory.

The Brush Company

One of the many scientists determined to improve the arc lamp was the American Charles Francis Brush born in Euclid, Ohio in 1849 and awarded the Bachelors Degree in Chemistry at the University of Michigan in 1869. Although he was a chemist, Brush was an avid experimenter with electrical equipment and soon realised that there was a large potential market for the electric light, provided that the existing equipment could be improved.[1] By 1875 he had designed and built his first dynamo but then had to appeal to his friend George Stockley, vice president of the Cleveland Telegraph Supply Company, for financial help. Stockley was impressed with the dynamo and agreed to make and market the dynamo and any other electrical equipment Brush could design. Before long Brush produced an arc lamp which had vastly improved carbons and an accurate but simple regulator, and it was this arc lamp that was chosen by the Franklin Institute of Philadelphia when in 1877 they decided to conduct trials of dynamos. The trials lasted for several months and in 1878 it was at last announced that the Brush machine was the best. The tests had been so thorough that one of the manufacturers withdrew his machine from the market, but to Brush the thoroughness had been a great advantage. The amount of information produced by the tests enabled him to make even more improvements and besides, the publicity he had received enabled him to sell his equipment to a wider public.

However, Brush felt that rather than supplying individual premises with dynamos he would be better supplying electricity from a central generating station; in that way he could also provide electricity for street lighting. With this end in mind, the California Electric Light Company of San Francisco was formed in 1879 – the world's first supply company with a central station – initially supplying twenty-two arc lamps in private premises from two dynamos. Within six months he was also supplying street lamps in the area and the number of arc lamps had grown to fifty.

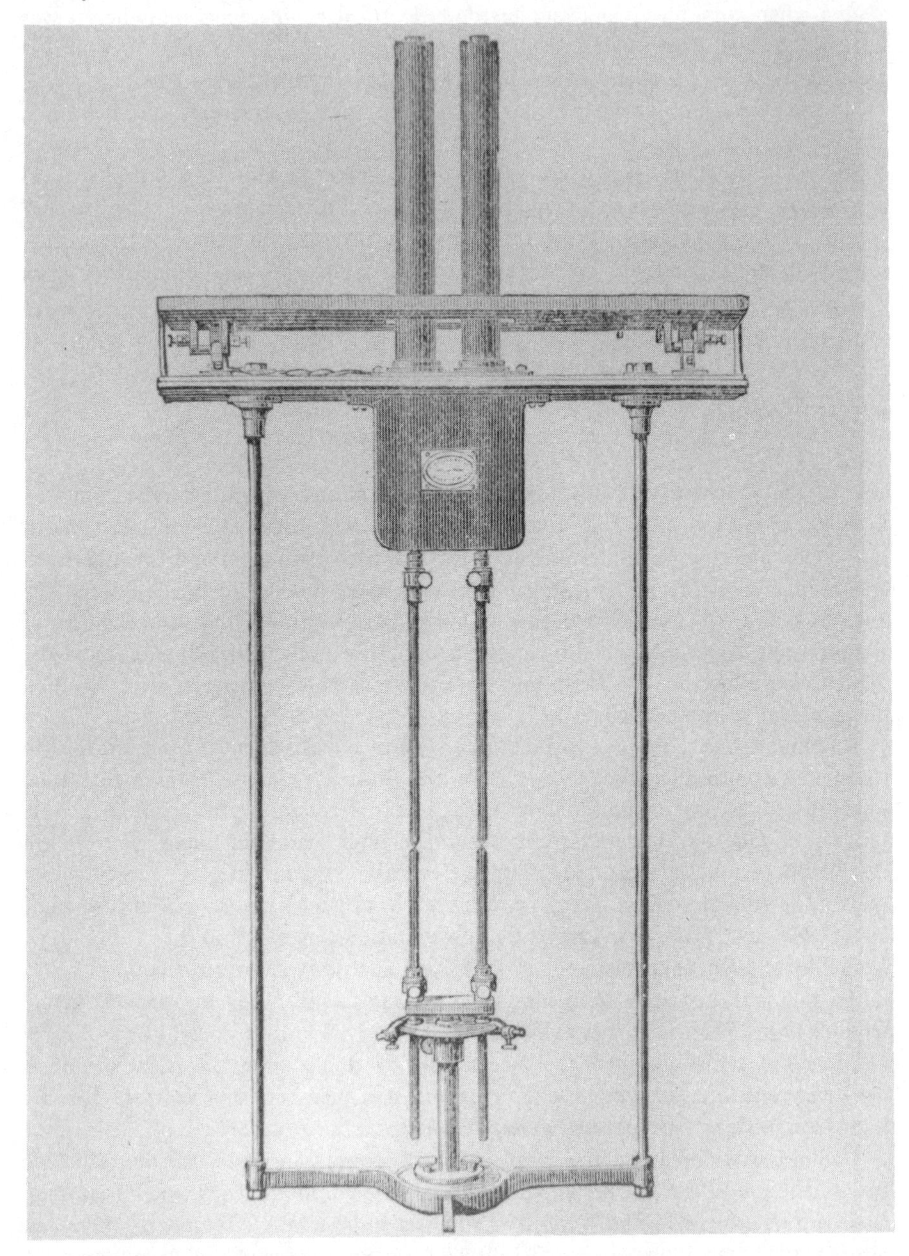

Fig. 10. *Brush double-carbon arc lamp ('Electric Lighting: Report of Joint Committee, Leeds 1882)*

From the beginning, the Brush Company was successful and soon spread across America and Europe, their strength being that they supplied reliable equipment which was simple to use and simple to look after. Simplicity was very important at a time when very few people understood electrical engineering, and when it was difficult to find anyone able to maintain or repair electrical equipment. This was particularly true in England where there were very few central stations until about 1890, most people having to look after their own installations on their own premises.

In America, Brush expanded by enlisting capital in cities and forming local companies, which although independent to a certain degree were obliged to buy equipment from the parent Brush Company who also claimed a share of the profits. Expansion in England was the same, the sale of concessions and equipment being via the Anglo-American Brush Corporation who held the English rights.

Robert Hammond

One of the purchasers of Brush concessions in England was Robert Hammond of London, who between 1 July and 21 December 1881 acquired exclusive rights to sell, erect and use Brush electrical apparatus in Northumberland, Cumberland, Westmoreland, Durham, Lancashire, Derby, Sussex, the Isle of Wight, Hampshire and Yorkshire. He also had a licence to sell and use Brush equipment in the City of Westminster. These rights and concessions cost Hammond £20,500 and included a 20 per cent allowance by Brush on net tariff rates, and the right to use Lane–Fox incandescent lamps.[2]

Hammond was more than just a business man, he was also one of the few men in the country with an unshakable belief in the future of electricity. Like Brush, he knew the future lay in the building of large central generating stations. In fact, he felt certain one day power stations would be built near coal mines, as it would eventually be much cheaper to transmit electricity long distances to the customer than to carry the coal across the country to local power stations. And how right he was! Today our modern coal-fired power stations are sited on the coal fields of Yorkshire and Nottinghamshire, and send their electricity all over the country.

To publicise his views (and his business) Hammond toured the country giving lectures about electricity, appearing at the Albert Hall in Leeds on Friday, 20 April 1883 before a distinguished audience headed by the Mayor, the High Sheriff of Yorkshire and his Chaplain, the Town Clerk and members of the Leeds Electric Light Committee.[3] The lectures were published as a book in 1883.[4]

Hammond also realised the importance of training engineers to look after his equipment and in 1882 he set up an Electrical Engineering College, which according to an advertisement in *The Times* of 2 May was to be set up in temporary premises, initially, at 2 Red Lion Square. 'To provide for the training of young gentlemen at Electrical Engineering,' *The Times* explained, 'the Hammond Electric Light and Power Supply Co. Ltd. have decided to take a limited number of PREMIUM

APPRENTICES, and, in order to give them a systematic instruction in Electrical Engineering, they have started the above college, containing complete laboratory and apparatus of the various systems.'

The short-term aim of Hammond and his company was to sell and install the Brush electric light, although he was willing occasionally to give demonstrations or to hire equipment out for special occasions. This was how the Brush lights had been made available during Gladstone's visit to Leeds. His ambition, though, was to provide electricity from a central station, and his first chance came at Chesterfield in 1881, when the town suddenly found itself without street lighting.[5] The contract with the gas company had expired and negotiations only resulted in the gas being cut off, so that on 1 September the town was in darkness. Hammond was given a contract to light the town by electricity for a trial period of three weeks, which was successful enough to be extended to a year. (A similar eight day experiment took place at Barnsley in December 1881, not as successful as at Chesterfield because, the *Leeds Mercury* reported, 'there were grave suspicions of foul play' which caused engine irregularities.)[6]

His most successful large-scale installation was that at Brighton, which began as an exhibition of arc lighting, also in December 1881. A central station was put up feeding an overhead system to which private customers were invited to connect, for the payment of 12/– per lamp per week. By mid-January the network was one and three-quarters miles long and a month later it was decided to make the installation permanent. All public supply systems in England up to that date had either already collapsed, or were soon to do so, but the Brighton installation survived through various owners and technical changes, and survives to this day in its modern nationalised form. It was thus the first continually surviving public system in England.[7]

Despite this success, though, large-scale supply was proving difficult as corporations were reluctant to go ahead until they had witnessed a satisfactory trial in their area, a business which was likely to prove costly in the short term but with the promise of eventual long-term profits. Hammond did not have the capital to undertake all the necessary trials over the large concessionary area that he held, especially now that he had also acquired the Brush rights for Warwickshire, Staffordshire and Worcestershire.[8] He had no alternative but to ask for public assistance and on 21 January 1882 a prospectus was issued for the formation of the Hammond Electric Light and Power Supply Co. Ltd.[9]

Initially, 50,000 £5 shares were offered of which 7,000 were retained fully-paid-up by Hammond who also took twenty of the shares as founders shares, which were entitled to two-fifths of the profit remaining after 10 per cent had been transferred to reserves and 10 per cent to shareholders. In other words, these shares attracted 32 per cent of the profits, a common method then of ensuring that founders of companies received extra payments. In fact Hammond expected these payments to be sufficient, for although he was the company's Managing Director he was to receive no other remuneration.

The statutory company meeting on 18 May 1882[10] confirmed the terms of the prospectus and further agreed to set up local independent 'Brush' companies, one of which thus formed being the Yorkshire Brush Electric Light and Power Company Ltd. The subscription list for this company opened on 19 May 1882, 100,000 £2 shares being on offer. Hammonds were to receive 25,000 shares (fully-paid-up) and £50,000 in cash in return for the Brush concessions for the whole of Yorkshire (plus the Lane–Fox incandescent lamps) and Yorkshire Brush were also to take over the staff and premises of Hammonds at 1 East Parade, Leeds.[11] So was formed the first major electrical company associated with Leeds; let there be no confusion though – despite its name and local office, this was quite definitely a London company with London directors.

The new company was established in a spirit of considerable optimism for Brush equipment had a reputation for quality, economy and reliability, and the introduction of the incandescent lamp (by Swan and Edison) had given new hope and impetus to the electricity industry as a whole.

But in a strange way the new lamps also proved to be something of a mixed blessing in some quarters, for many manufacturers discovered that their dynamos were not suitable for supplying incandescent lamp installations. The problem was that series-wound dynamos had been designed to give constant current, in order to best supply arc lights. For multi-lamp installations of the incandescent lamp, however, the load fluctuated as lamps were switched on and off and a machine giving out a constant current soon burned out the bulbs as the installation was off-loaded.

The Ferranti dynamo

The failure of Brush – in particular – to produce a suitable range of variable current machines was both a surprise and a blow to the Hammond Company and its subsidiaries, dependent as they were on Brush equipment. Robert Hammond was well aware that the future lay with incandescent lighting and his concern at Brush's failure was lessened somewhat when he heard of a new dynamo which had recently been designed by a young engineer, S. Z. Ferranti.

Hammond no doubt heard about the new machine from his solicitor, Francis Ince, who was one of the partners in the company which had been established to make and sell the dynamo. Hearing that the company had severe financial problems Hammond offered £5,000 of his own money and bound himself to a further £60,000 on condition that Hammonds should become sole agents for the Ferranti machine. This was an offer gladly accepted by the new Ferranti company.[12]

The Ferranti machine was not just a dynamo capable of supplying incandescent lighting; it was also a considerable advance on existing dynamos, its particular advantages being its simplicity, its small size, and its light weight.[13] Public interest was greatly excited by the various reports which circulated, but the introduction of the machine to the public at the beginning of December 1882 did not have the depressing effect on other companies' shares that had been expected. In fact, in a curious way it had a detrimental effect on some of the companies it was supposed to assist, being a cause of puzzlement, for example, at a special meeting of the Yorkshire Brush Company held on 15 November 1882.[14] The board had decided to convene the meeting as a result of two circulars issued by a shareholder, Mr. Whiteley, who made such serious allegations that a decision had to be made either to continue the existence of the company or agree to its immediate dissolution.

One circular had informed the shareholders that the licence for the Lane–Fox incandescent lamp was not, after all, an exclusive concession for the company's area, but merely a licence to use the lamp, a licence shared by other companies. The board explained that the language of the licences had appeared to the directors then, and did so now, to be direct and exclusive, and several other companies had thought the same. No doubt or question on this subject had arisen until Sir Charles Bright wrote, on 3 October, and pointed out that there was a concurrent licence granted to the British Electric Light Company. The directors were of the opinion however that the Lane–Fox lamp now being manufactured by Brush was so different from the original that the concurrent licence was irrelevant.

Turbulent times

Mr. Whiteley also claimed certain irregularities in the setting up of the company, in that not enough shares had been taken to satisfy the terms of the prospectus, causing a serious deficiency of capital. After payment for concessions (the Lane–Fox lamp included) only £23,266 was available to work an area described in the prospectus as one-ninth the whole of England. The concessions had cost £100,000; Mr. Whiteley said he had read of Mr. Hammond's statement concerning the Ferranti machine and wished to know if this was to supersede the Brush machine. If so, what had they paid their money for?

Mr. Lowe, presiding, said that a forty lamp job at Middlesborough was nearing completion and a repeat order was expected. Expenses had been kept down and they had given up the idea of exhibiting or doing any free work. They might not make sufficient money to pay a dividend at the end of the year, but they could not expect rapid progress against the enormous monopoly of the gas industry. The company was making fair progress at little cost. If it was wound up, the enormous charges of the lawyers and liquidators would leave the shareholders nothing but ashes. When put to the vote, the winding-up resolution was opposed and the directors were empowered to carry on the business.

A short time later, on 19 January 1883, an Extraordinary General Meeting was held, this time 'for the purpose of considering an offer for the re-purchase from this company of the Lane–Fox licence, for the sum of £35,000, payable in fully-paid-up shares of this company, and if then considered desirable, passing resolutions authorising the board of this company to carry out a sale of the said licence upon the above terms, or such other terms as the meeting may decide'.[15] The meeting was told that the value of the Lane–Fox lamp to the company was nil: under the new Electric Lighting Act customers could choose any lamp they wished, yet under the terms of the overall concessions they were bound to the Lane–Fox lamp. Confusion arose, though, as to who was actually buying back the licence, the Electric Works Co. being quoted, although reportedly going into liquidation. They had evidently acted as brokers between Hammonds and Yorkshire Brush on the

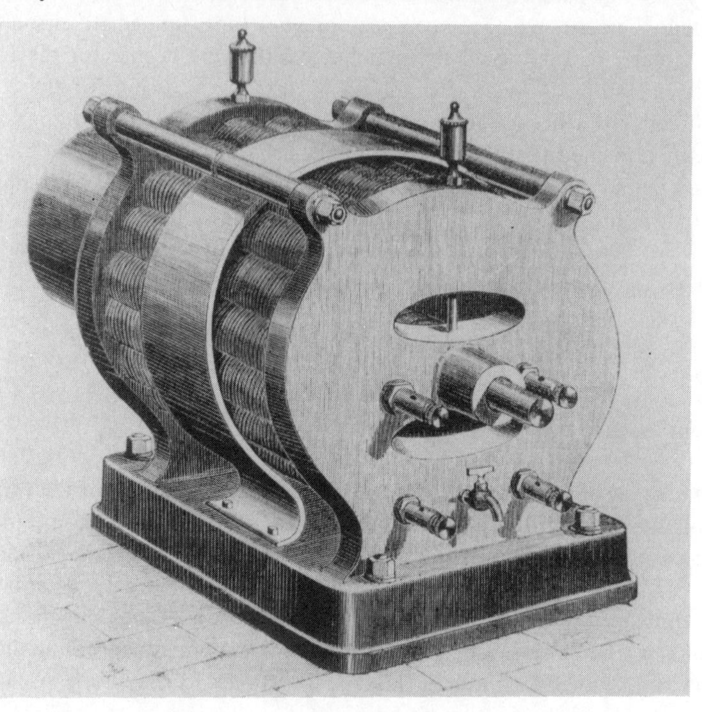

Fig. 11. *Ferranti dynamo (*Electrician, *2 December 1883)*

formation of the latter company, but Mr. Hammond assured the meeting that in effect Hammonds would be making the payment. A solicitor for Electric Works Co. eventually altered the offer to £30,000 shares plus £5,000 in cash, an offer which was accepted by the meeting.

While Yorkshire Brush were riding out this turbulent time, Hammonds were having storms of their own to weather. For them, 1882 was, like the curate's egg, good in parts. The *Electrician*, in a review of 1882,[16] reported that the installations at Brighton and Chesterfield were still running successfully, although the original Lane–Fox lamps at Chesterfield had been universally condemned and had been replaced by the new type. Similarly, the lighting of the Mersey Tunnel had been disappointing due to the Lane–Fox lamps. Except for these problems, though, their installation work had made steady progress, two more town installations having been made at Cockermouth and Portsmouth and several existing customers adding to their lights. The real problems for the company had come from the unlikeliest of sources: Brush, the parent company. As Robert Hammond explained

Fig. 12. *Ferranti dynamo – carcass (*Electrician, *3 November 1883)*

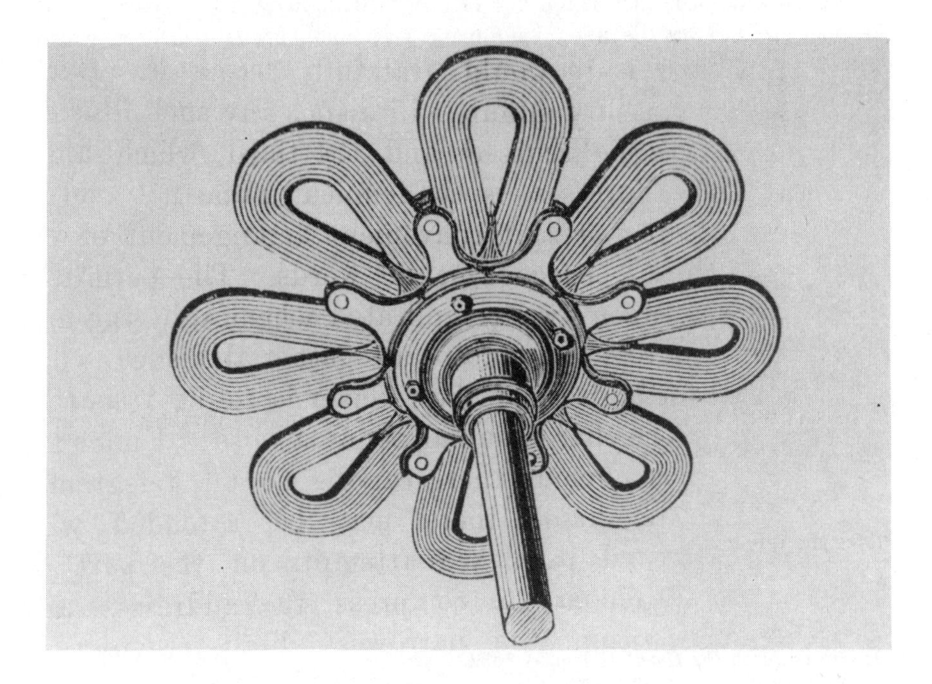

Fig. 13. *Ferranti dynamo – armature (*Electrician, *3 November 1883)*

at the company's second ordinary general meeting on 9 May 1883, litigation was proceeding over royalties and deliveries of goods, and it was evident that they could not continue under the banner of the Brush Company who were being unreasonable on two counts.[17]

First, the Crystal Palace exhibition at the beginning of 1883 had been disastrous to Brush. It had publicly been seen that Brush was not the only company in the field and should therefore reduce prices. But, suicidally, Brush had declared a dividend of 100 per cent (not a return of capital, but a dividend!) which had only convinced the public of the company's irresponsibility.

The other cause of dissatisfaction with Brush was the company's inability to produce a large machine for incandescent lamps. Robert Hammond had already told the Yorkshire Brush shareholders that it was no use going to a manufactory with 5,000 gas jets and telling the proprietor they had a machine which would run only 80 lights. What was required were machines capable of running 500, 1,000 or 5,000 lights. When Brush had been asked for a 500 light machine, they had offered two 200 and one 100 light machines, at the ridiculous price of £1,800. This was the reason for taking on the Ferranti machine, an ideal opportunity to become independent of Brush and have the whole field to themselves over the whole country.

To increase this independence from Brush, Hammonds also paid £2,878 to take over the patents of the incandescent lamps made by Wright & Mackie, two young electricians who made their lamps in a factory in Bermonsey. The amazing part of the manufacture was the use of automatic glass-bulb blowing machines, which enabled young lads of fourteen or fifteen to turn out 300 good bulbs a day.[18] Previously, highly skilled and highly paid glass blowers had been required, producing a fraction of this number. The double spiral filaments gave out a good light and Hammonds hoped that this combination of high technology and mass production would help them in their aim of independence from Brush.

Although a 5 per cent dividend was announced in May 1883, the year had been a struggle and Robert Hammond claimed his constant attention to the company had been required, a circumstance not expected when the terms of his employment had been negotiated. The shareholders therefore agreed to the proposal to pay him a salary of £1,000 a year (with £500 for the year just ended).[19]

Robert Hammond must have been disappointed that his company was not prospering sufficiently to provide him with a reasonable income from his founder's shares; business was not good.

Notes and references

1. The story of Brush comes from *The Electricity Manufacturers 1875–1900*, H. C. Passer, (1953) pp. 14–21.
2. Prospectus in *The Times*, 21 January 1882.
3. *Yorkshire Post*, 21 April 1883.
4. Review of *The Electric Light in our homes* (London: Warne & Co.), in *Electrician*, 19 April 1884. Also – Obituary in *Electrician*, 6 August 1915.

5. *Engineering*, 14 April 1882.
6. *Leeds Mercury*, 10 December 1881.
7. *The Early Days of the Power Station Industry*, R. H. Parsons, (1940).
8. *Electrician*, 20 May 1882.
9. Prospectus in *The Times*, 21 January 1882.
10. *Electrician*, 20 May 1882.
11. *Leeds Mercury* 22 May 1883.
12. *Electrician*, 12 May 1883.
13. *Ibid*, 2 December 1882.
14. *Ibid*, 18 November 1882.
15. *Ibid*, 27 January 1883
16. *Ibid*, 6 January 1883.
17. *Ibid*, 12 May 1883.
18. *Yorkshire Post*, 19 May 1883.
19. *Electrician*, 12 May 1883.

Chapter 4

Gas: a stern rival

With the introduction of the incandescent lamp in 1879–80 there was a hope and a belief that the popularity and the application of the electric light could now spread more rapidly. Alas, in the seventy years since Davy had discovered the arc there had arisen a much sterner rival to the electric light than mere oil lamps and wax candles – the gas light was now a well established and popular system of lighting.

The first practical coal gas had been produced in 1792 by Murdoch, an employee of the Birmingham engineers Boulton and Watt. Ten years later the Boulton and Watt Soho factory in Birmingham was lit by gas and was probably the first place so lit. The gas light at that time (and indeed for the next eighty-three years until the invention of the gas mantle) consisted simply of holes in a tube through which the gas was fed and ignited to burn with a pale yellowish flame. The only real improvements in those early years concerned the concentrating of the lights of several gas flames by so shaping the holes that the flames burnt in the shape of a fan or a batswing. Even with these improvements, however, the gas light – though brighter and more convenient than candles – would have seemed very dull to our modern eyes. We must remember that really bright gas lighting was not available until Karl Auer of Welsbach in Austria invented the Auer incandescent gas mantle in 1885: even then the new light was not generally available for several years. The mantle was apparently first available in Leeds as late as 1899 when the Welsbach Incandescent Gas Light Company of 158 Briggate, Leeds, was able to announce that their light was 'now for sale'.[1]

Benefits and dangers: gas versus electricity

Even without the mantle, however, the gas light was such an improvement over previous means of artificial lighting that it rapidly gained in popularity and by the time the incandescent electric lamp was introduced gas lighting was in general use, even in the homes of the poor. Despite this popularity, though, it was always recognised that there were certain disadvantages, dangers even, in the use of gas and the availability of the new electric lamp provoked new discussions on the relative merits of the two systems.

The main objections to gas, it was argued, was that to provide light it had to burn, thus giving off heat, consuming the oxygen of the atmosphere and giving off chemical effluence and soot. There was also an everpresent danger of fire and explosion.

One of the most obvious results of these defects was that gas lighting needed a good supply of air to allow it to burn; a Victorian drawing room lit by gas was of necessity very draughty, with fresh cold air constantly being drawn into the room to feed the gas flame. And yet the gas provided heat, so that in the upper parts of the room the air was very warm and lacking in oxygen. The atmosphere at head level could be overpowering and it's not surprising that ladies of the nineteenth century had a reputation for frequent bouts of fainting and giddiness.

The products of combustion included carbonic acid gas, sulphuric acid, and soot so that rooms which were gas lit needed constant re-decoration, because of the soot, and frequent renewal of fabrics and furnishings, because of the deleterious effects of the acid effluent. In some circumstances this was of such importance that gas was excluded from the premises. Libraries, for instance, could ill afford to suffer major damage to their leather bound books (leather was particularly susceptible to damage) and annual re-decoration could often be a major item of expenditure, particularly when scaffolding was required. In many cases gas was also excluded from premises because of the fire risk (in furniture factories for instance), or if gas could not be excluded – because artificial light was necessary – the risk of fire was grudgingly accepted in the knowledge that tragedies could, and would, occur. Theatres and public halls, for example, needed artificial light, but candles and later, the gas light, were recognised as a major cause of fires and disasters, the naked flames on and around the stage being particularly dangerous due to the proximity of inflammable costumes and scenery. There was not always enough care taken: 'We who are "behind the scenes"', wrote a theatre worker, 'soon get used to the reckless manner in which the gas and lights are used on the stage, but whenever friends come "behind" to see me during a performance they invariably express their astonishment at the alarming manner in which the lighting is carried out.'[2]

The electric light, generally speaking, had none of these inherent defects of gas lighting – it was clean and safe, and the light it gave out was brighter and softly white, so that rooms lighted by electricity were cool and comfortable, with a pleasant light more approaching the effect of daylight than the yellow light of gas.

These properties were uniquely demonstrated together at the Savoy Theatre in London on 28 December 1881, when for the first time in any theatre gas was abolished from the stage and replaced by Swan incandescent electric light (which had already been installed in the auditorium since the opening of the theatre). Advertisements for the Savoy Theatre pointed out that 'the air in all parts of the house, including the gallery, is perfectly pure during the entire performance'.[3] The advertisements also stressed the safety of the new means of illumination, quoting, apparently, from the *Daily News*: 'A herd of wild buffaloes suddenly turned loose among the intricate details of the fine woodland scene of the second act, of PATIENCE, might trample down "groundrows" and sweep away, and destroy whole "battens" and hanging "wing lights", scattering literally hundreds of lamps among the wreck of inflammable canvas and frail woodwork, but assuredly they could not set fire to the Savoy stage while lighted only by this means.'

To prove this safety to the audience, the manager (Mr. D'Oyly Carte) on the first night covered one of the Swan lamps with a piece of highly inflammable muslin and then broke the glass. The incandescence in the lamp was extinguished immediately the vacuum was broken and the muslin was completely unmarked.

The light from the bulbs was also considered to be a great success, the effect being 'pictorially superior to gas, the colours of the dresses – an important element in the "aesthetic" opera – appearing as true and distinct as by daylight'.[4] The benefits of the electric light were neatly summed up by the *Telegraphic Journal and Electrical Review*: 'The advantage to audience and performers is very great as regards the moderated temperature of the Theatre, the absence of noxious fumes and the reduced risk of fire.'[5]

Many of these allegations were, of course, strongly refuted by the gas industry, who claimed that the electric light was not as good as the publicity made out, and it was indeed true that defects in the electric light could easily be found. Arc lights, for example, were themselves a recognised fire hazard, due to the likelihood of red hot carbon or molten copper falling from the burning rods. Many fires were caused in this way until the design of the glass globes was improved to catch the hot pieces before they fell to the floor.

There was danger, too, in the early practice of using bare rather than insulated wires in electric lighting installations. It was quite possible for these to be inadvertently shorted together, through either damp or dust, thus causing a short circuit current and the possibility of fire. The short circuit current could also overheat the wires themselves if not switched off quickly, and red hot wires or damage to the generators themselves was a common event until Edison introduced the fuse, which safely burnt away when excess current flowed and so cut off supply. Bare wires were also a danger to untrained workmen who could easily be electrocuted. This was not really likely with incandescent lighting installations, which usually ran at about 200 V, but arc lighting systems could run at a thousand volts or more and there was a serious risk of death if the wires were touched.

These problems were, without doubt, serious, but they were purely problems of design or installation. Gas was a potentially inflammable and explosive fuel, no matter how well installed, whereas electricity was only dangerous if the design or installation were inadequate or the workmanship was at fault. Those who had knowledge of such things felt that such incidents that did occur were altogether inexcusable, but it was difficult to find electricians with enough experience or training to work safely on what was, after all, a completely new technology. The magazine *Engineering* was quite adamant about this. 'There cannot exist the very smallest fear of fire occurring in an installation of incandescent electric lighting, if the conductors are properly constructed, and put up by a person who understands his business; and the same remark applies with equal force to the question of the danger of electric shocks.'[6]

There seemed little doubt, from the evidence presented, that the electric light was far superior to the gas, provided it was properly designed and installed. At the present stage of development the electric light could only get better, whereas gas could never overcome its inherent defects of effluence, explosiveness and ennervation. The public began to anticipate with delight the spread of the electric light.

Worst gas in Britain

In Leeds the prospect of the installation of the electric light on a wide scale was viewed with even more eagerness than it was in other parts of the country, for Leeds, unfortunately, had the reputation (among the townsfolk at least) of having the worst gas in Britain. It was extremely dirty, and there were allegations also that the pressure was low and that the illuminating power was poor.

Gas was dirty at the best of times, of course, but in Leeds there were particular troubles due to the large amount of sulphur in the gas, a problem which the Gas Committee tried to cure by investing in new gas-making apparatus and processes, but with little success. This sulphur, after passing through the gas light, eventually turned into oil of vitriol – sulphuric acid – which had a very harmful effect of furnishings, particularly leather. J. D. Heaton, of Claremont, was of the opinion that the Leeds gas contained 'more sulphur than that of most other large towns'. So severe was the effect of the acid in his house that he had been forced to restrict the use of gas at home. He had 'for some years returned to the more primitive use of oil lamps'.[7]

The *Leeds Mercury* had already complained earlier that year (1879) of both the offensive effluvia and the low pressure ('in the dwellings of the humbler classes there is absolutely little or no pressure of gas')[8] and the Leeds Gas Engineer was eventually tempted to defend his gas undertaking. This did little but draw this riposte from Jackdaw: 'I admire "pluck", and therefore I cannot pretend to be dissatisfied with the letter of Mr. Woodall, the Leeds Gas Engineer. This gentleman bravely contends that he does the public of Leeds a great service by supplying them with impure gas. The gas smells so badly when it is allowed to accumulate in a room, that Mr. Woodall contends it acts as a kind of danger signal, and warns you that you ought to look after your ventilators.'[9]

Complaints continued over the years and seemed to reach another crisis in the winters of 1885 and 1886. In a letter to the *Yorkshire Post* 'Benighted', for example, described gas street lamps which he said shed forth 'visible darkness'.[10] 'Rushlight', a few months later, complained of being 'blinded by the poor quality of the gas and almost suffocated by the intolerable stench emitted by the abominable mixture supplied to Leeds consumers under the name of gas.'[11] In November 'Corrigenda' commented bitterly that 'the Leeds gas is bad; worse than it has been known to be for many years. The lights are dull, the impurities from combustion are in excess of what they should be, and the gas is mixed with air.'[12] Shortly after this, a 'Market Man' complained of the gas in the Leeds Covered Market, referring in his letter to the 'wretched and inadequate lighting.'[13]

There is obviously little doubt that the gas lighting in Leeds was of poor quality, but strangely the fact that it was 'dull' and 'wretched' was never officially recognized. The regular reports of the tests for candle power at the works gave no indication of deficiencies – but according to allegations made in a letter from 'Rushlight' this was not surprising, for the tests had never been properly conducted. In his letter he explained that by statute the Leeds gas had to be tested using a 15-hole Argand burner and should have given a light equal in intensity to 16 sperm candles of six to the pound, burning at the rate of 120 grains per hour. Instead, however, a 24-hole burner had been used and a monthly assurance had been given that the gas was equivalent to 16 candles; the 24-hole light was naturally much brighter than a 15-hole burner and the true illuminance was really much nearer to 14½ candles, well below the statutory minimum. According to Rushlight, the magistrates inspector had carried out independent tests in 1881 and at a time when the official tests showed an average illuminating power for a month (in August) of 16.69 candles, the independent tests had showed an average of only 15.95. Moreover, the gas had been below standard thirteen days during the month, sometimes to the extent of two candles, and on one occasion it had been found to be no less than five candles below the statutory standard.

The local magistrates, who in theory had the power to monitor and control the illuminating power of the Leeds gas, had a similar problem in 1887, discovering that the actual illuminance was less than both the declared and the statutory figures. Their disputes with the Town Clerk (for the Gas Committee) was revealed in the *Leeds Evening Express* of 28 October 1887 which gave details of the correspondence which had taken place over the last few months. This had started in July 1887, when the Justices measurements at Dewsbury Road (supplied by gas from Meadow Lane) had shown values frequently below 16 candles in the month despite an official average for the month at Meadow Lane works of 17.05 candles. This discrepancy had caused the Clerk to the Justices to enquire whether the testing apparatus used by the Gas Committee was defective, in either working or location, and if in fact the apparatus satisfied the requirements of the Leeds Corporation Gas Act of 1870 by being 'so placed as to afford a test for *all* the gas supplied by the Corporation'.

Eventually, towards the end of September, the Town Clerk replied that the photometer had been tested and was in good order, which caused the Justices to ask, of course, how the discrepancies could have been caused, if the apparatus was working properly. The Town Clerk replied that he was referring the matter to the Gas Committee, which was due to meet on 20 October; a week after this date – on 27 October – the Justices Clerk wrote again to the Town Clerk asking why there had been no response to the question of the discrepancies. The Town Clerk merely repeated, in reply, that the matter was being referred to the Gas Committee.

There was no further correspondence for a couple of months until 20 December, when the *Yorkshire Post* published a letter from the Town Clerk to the Justices Clerk, in which he expressed his regret that the affair had been publicised. He also said that there was at present no explanation of the discrepancies but pointed out that the Justices had no right to test the gas without a gas officer being present, should not have tested gas unless it was actually for consumption (some tests had been made when no gas was being supplied) and in any case had no legal right to test the gas at all. He suggested that a conference be arranged to set up an Inspector on a legal basis.

On Christmas Eve 1887 the reply from the Justices Clerk was printed in the *Yorkshire Post*. The Justices sternly reminded the Gas Committee that they did indeed have statutory powers which were capable of being called into exercise within an hour, authorising them to do what had hitherto been done, and more beside. They also warned that under these powers they could extract a penalty of £20 on each occasion a test witnessed by a gas officer showed the gas to be below 16 candles. However, they had no wish to be severe in this matter and agreed to the conference.

Discussions were held during January 1888 and the chairman of the Gas Committee – in conciliatory mood – finally agreed to the demands of the Justices: the tests would be done at places from where the gas was supplied (rather than at the works), there would be present at the tests a Gas Committee representative, and there would be an improvement in the illuminating power of the gas. The agreement was ratified by the full Council on 8 February 1888.[14]

It's not clear whether there was actually any improvement in the power of the gas over the next year or so, for there were conflicting opinions. On the one hand the chairman of the Gas Committee asserted strongly at the beginning of 1889[15] that the average illuminance had been well above the standard; on the other hand there was an adamant allegation from Ald. Spark (who had previously served on the Gas Committee) that the gas had never been up to the standard required – as compared with 1884 the gas for the second half of 1888 had been not less than one candle below that supplied in that year. More seriously, he was of the opinion that even this low figure was greater than the true illuminating power, for he could prove that there was available at the gas works a special pipe of high-quality gas which was used for the testing.[16] The Gas Committee must surely have wished that these doubtful or unsubstantial allegations about the power of the gas would be the full extent of their problems that year, but unfortunately, however, they had other difficulties to contend with in 1888.

First there came serious claims of fraud, made by a Mr. Ellis Lever of Colwyn Bay, a coal dealer of high standing and reputation. He had recently exposed frauds in the gas undertaking of Salford and now he was insistent that similar frauds were taking place in Leeds. His claim was that inferior coal was being provided by the collieries, that the Gas Committee knew this and yet no financial adjustment was being made. Such was his reputation and his insistence that in the autumn a series of tests were arranged.

The results of the investigation were discussed at a Council meeting on 14 November when it was admitted that no defect in the quality of the coal had been discovered.[17] It had been found, however, that official chemical analysis of coal samples was only done half a dozen times a year and that 28 per cent of these few samples were below par. It was admitted that although no fraud had been proved 'there had been very great room for frauds to be practised'.

In a letter to the *Yorkshire Post* on 10 January 1899 Mr. Lever stated that it was not surprising that the tests had found no defect, for the simple reason that the samples tested at Meadow Lane on 7 November were not the samples he had bagged on 24 October. Despite his protests he had not been allowed to repeat the tests. He also pointed out that chemical make-up was not the only measure of quality. The coal contracts, he explained, expressly stipulated that the coal should be free from base, shale, pyrites, slack, bind and hub and yet the trucks he had inspected were full of pyrites, slack and hub. Mr. Hunt (chairman of the York Street Works sub-committee) had exclaimed that the trucks contained the worst coal he had ever seen.

Despite the fact that this recognition of poor quality had been made in the presence of the Mayor and the Town Clerk there was apparently no reference to the defects in the Council's report. The matter was therefore discussed again by the full Council on 6 March 1889,[18] when an analysis of the previous five-year figures of coal quality was presented. To the amazement of some, the Gas Committee's own figures showed that out of more than 1 million tons of coal purchased in that period, almost 600,000 tons were below standard. Because of the lack of financial compensation there had been a loss of £14,744, or nearly £3,000 a year. There was naturally some degree of public suspicion which was aggravated when it was revealed that several of the Gas Committee members were also proprietors or directors of the collieries which had supplied the coal,[19] but the Council accepted the explanation that, although there had been an apparent loss due to inferior coal there had also been a corresponding gain because the good coal was much higher than the standard.

During all this investigation and accusation another different event was brought to light, coming to public attention during discussion of Gas Committee minutes by the Town Council at its meeting on 2 January 1889.[20] Here it was revealed that in December a hole had been discovered in the top of No. 5 gasholder at York Street Works. The ex-manager of the works had been written to at his new place of employment in Bristol asking for his explanation but he had returned a 'studiously insulting' letter which offered no immediate reasons. The Council were unwilling to attempt to solve the mystery, but to Ald. Spark there was no mystery. It seemed obvious to him that in response to the demand made upon the manager to produce large quantities of gas he had been forced to produce large quantities of bad gas, which after registering he had passed into No. 5 gasholder. From there it had been allowed to vent through the hole, thus preventing the passage of the bad gas to the customers.

When all things are considered, it is hard to disagree with Ald. Spark's opinion that there was 'a vast amount of trickery going on in the gasworks of the borough of Leeds', [21] although there was as yet no definite proof of fraud or mismanagement, nor had the Leeds townspeople suffered directly from any of these incidents. It was not to be long, however, before the whole of Leeds was directly involved.

Gas dispute

Within eighteen months the members of the Gas Committee found themselves in dispute with the stokers in the gas works. The dispute proved to be incapable of resolution and on Monday 1 July 1890 all the gas workers went on strike. Despite the importation of 'strangers' from London and Manchester there was a progressive loss of gas in Leeds so that on that first day there was what was described as a 'new experience' in the town[22] – Leeds fell into total darkness.

The use of 'strangers' soon provoked violence and riots so that the police and the dragoons had to be called out and the riot act read. Many people received severe injuries. It's not surprising that many of the 'strangers' were persuaded to leave their work.

For several days there was no gas throughout the town, and 10,000 people were made idle as industry closed down. It was this lack of gas that caused the *Yorkshire Post* works and offices to have electric light installed for the first time. Initially, the paper was produced with the aid of wax candles, but on Tuesday 2 July at 11 am Mr. Hartnell (a Leeds electricial engineer) received the order for the installation of the equipment. The light was at work just after 11 o'clock that same night. The engines used to drive the printing machinery were adapted to drive a 190 16-candle dynamo running at 2,000 rpm, supplying electricity to forty 50-candle incandescent lamps, twenty 16-candle lamps and six 100-candle power lamps.[23] In the *Leeds Mercury* offices, incidentally, the electric light had been in use 'for some time.'[24]

The strike was finally settled through the Chamber of Commerce, the men returning in the afternoon of Friday 4 July. It was fortunate that the Chamber of Commerce had been both willing and able to mediate in the dispute, for the Gas Committee had 'stood by in craven fear, impotent in the storm, paralysed – contemptible in the eyes of the law abiding.'[25]

To most people the settlement of the strike and the return to work was the end of the matter, but not apparently to the Gas Committee, who took the earliest opportunity to try to get rid of those they considered the leaders of the strike. Because of the seasonal demand on the gas system it was quite normal to give notice to many of the men employed in gas production at the end of the winter, indeed many men took the job on the understanding that it was only for the winter months. In January 1891, though, in contravention of normal custom and practice the sixty men given notice were not short servers but the longest serving men,[26] many of whom had indeed been leaders of the strike. A second strike was immediately threatened, and was only averted when the Gas Committee agreed completely to the men's terms.[27]

The strike – and the threat of a strike – was undoubtedly a cause of great inconvenience to the Leeds public and a cause of hardship, too, to those who found themselves out of work. But to most people it was also further proof of the total inadequacy of the Gas Committee and of the gas supplied from the Leeds Gasworks.

Electricity: superior but too costly

Yet despite all the complaints, and despite the acknowledged superiority of the electric light over gas, there was still no widespread installation of the electric light in Leeds, even as late in the century as 1891. How could this be? The answer, of course, was cost!

There was no doubt that for a small house the capital cost of supplying a boiler, steam engine, dynamo and lighting system far outweighed any benefits the electric light might have over gas. Boilers and steam engines were not cheap, of course, although readily available, and electrical equipment, still being developed, was very expensive. Dynamos, for example, could cost anything from £75 for a one-arc light model to £720 for a Brush 40-light machine. Arc lamps themselves were priced at £10 each and incandescent lamps ranged in price from 5/– to 12/6.[28] These prices represented large proportions of personal income and far exceeded many people's wages at a time when such worthies as the Chief Inspector of Waterworks in Leeds was receiving a salary of only £175 p.a.[29] and the Borough Engineer had an annual salary of £600.[30] Furthermore, the equipment required the provision of special premises – a boiler house or engine room – and a boilerman or engineman had to be available. On the considerations of capital and accommodation alone, therefore, it was evident that the electric light could only be considered for large industrial or commercial premises, probably where there was already a boiler house and steam engine.

Running costs also needed consideration but these were much more difficult to evaluate. Despite various controlled tests and exhibitions by many of the electric light companies, assertions of costs proved difficult to substantiate. All it was possible to say with accuracy was that arc lamps gave a lot of light at little cost, whereas incandescent lamps were more expensive but were more flexible in their application and more suitable for smaller rooms. Comparison of both systems with gas was made difficult by the fact that the price of gas varied widely throughout the country.

Surprisingly, perhaps, the country's cheapest gas was to be found in Leeds, the price in 1880 being 1/10d per 1000 cu ft.[31] The Corporation had taken over the local gas undertakings from the Leeds Gaslight Company and the Leeds New Gas Company in 1870 at a cost of £763,225,[32] and under their control the price of gas fell drastically, due in no small measure to a policy of selling the gas at virtually cost price. The Council believed that the gas user should benefit from economies rather than the general ratepayer.

The Council's policy and achievement was laudable, of course, but there was some public cynicism about the low prices. Ald. Spark, for example, commented that the Leeds public would prefer an improved gas to a low price[33], and there was speculation that Leeds gas at 1/10d per 1000 cu ft was actually more expensive than other gas at 3/– per 1000.[34]

Despite these adverse comments, however, there could be no denying that the Leeds gas was cheap in price, a fact which made it especially difficult to justify the installation of the electric light on economic grounds. Certainly for private persons there was little financial incentive to invest in the new technology and it was only on aesthetic or safety grounds (where the virtue of the electric light was in itself desirable) that electricity could be justified. Nor was there any incentive for the Council to initiate electric light supplies, for by so doing they could only detract from their own gas undertaking and put at risk the huge investment they had made in the gas works. In 1880 at least there was no prospect of a widespread use of the Electric Light, not without a drastic decrease in its cost.

Notes and references

1. *Leeds Mercury*, 20 February 1899.
2. Letter in *The Times*, 30 December 1881.
3. *The Times*, 29 December 1881.
4. *Ibid*, 29 December 1881.
5. *Telegraphic Journal and Electrical Review*, 7 January 1882.
6. *Engineering*, 31 March 1882.
7. Letter in *Leeds Mercury*, 5 August 1879.
8. *Leeds Mercury*, 20 February 1879.
9. *Ibid* (Saturday Supplement), 22 November 1879.
10. *Yorkshire Post* (letter), 10 December 1885.
11. *Ibid* (letter), 5 August 1886.
12. *Ibid* (letter), 24 November 1886.
13. *Ibid* (letter), 3 December 1886.
14. *Ibid*, 9 February 1888.
15. *Ibid*, 4 April 1889.
16. *Ibid*, 7 March 1889.
17. *Ibid*, 15 November 1888.
18. *Ibid*, 7 March 1889.
19. *Ibid* (letter from 'Verax'), 3 November 1888, and (letter from E. Lever), 10 January 1889.
20. *Ibid*, 3 January 1889.
21. *Ibid*, 7 March 1889.
22. *Ibid*, 2 July 1890.
23. *Ibid*, 3 July 1890.
24. *Electrician* 4 July 1980.
25. *Yorkshire Post*, 27 October 1893.
26. *Ibid*, 6 February 1891.
27. *Ibid*, 24 February 1891.
28. 'Electric Lighting', Report of Leeds Joint Deputation, 1 May 1882.
29. *Leeds Mercury*, 12 August 1880.
30. *Ibid* (Saturday Supplement), 24 September 1881.
31. *Ibid*, 12 August 1880.
32. *Leeds Tercentenary Handbook*, 1926 – p. 113.
33. *Leeds Mercury*, 4 August 1879.
34. *Yorkshire Post*, 10 April 1889.

Chapter 5

Electric Lighting Act 1882

The major obstacle to a drastic reduction in the price of the electric light lay in the fact that electric light companies (or corporations wishing to supply the electric light) had no legal right to lay their cables in public streets. They were compelled instead to run their cables along private roads, railways or canals, or along the roofs or walls of private premises (provided they could get permission) and failing this they were restricted to supplying complete systems – including engine and dynamo – to individual customers.

Gas companies, on the other hand, did have the right to lay their pipes in public highways, by virtue of the provisions of their enabling Acts and therefore were able to provide supplies from central gas works. This gave economy of large scale production, a good utilisation of equipment and cheaper standing charges. Gas consumers were therefore able to avail themselves of a cheap fuel and also had no need to provide accommodation or money for their own plant.

This legal restriction was a handicap felt keenly by the electricity companies and the disadvantage was recognised by the Press. The *Leeds Mercury*, for example, commented that: 'until the electricity companies are placed on the same footing in this respect as the gas companies, we can have no satisfactory comparison of the relative cost of the two modes of lighting. So long as electric lighting is handicapped, directly or indirectly, so long will it be placed under a disadvantage as to cost.'[1]

It was only Parliamentary action which could change this state of affairs and on 13 April 1882 a new Bill was presented to Parliament designed to allow the establishment of central electricity undertakings with the same rights of excavation in public highways as the gas and water companies. There was a determination in this Bill, however, to prevent the granting of perpetual monopolies such as had occurred with these other service industries, for the public had grown to dislike their excessive charges and high-handed methods. The Bill proposed, therefore, that local authorities would control electricity supply by either undertaking the supply themselves or by overseeing the activities of companies within their area. To do this the Bill required that undertakers would be established by Licence (if approved by the local authority) or by Provisional Order, to be confirmed by Parliament, with the right of purchase of the undertaking by the local authority after a set number of years, at a price exclusive of goodwill.

On the same day that the Bill was presented to Parliament, the Gas Committee in Leeds met to consider it.[2] Both the Town Clerk (G. W. Morrison) and the Mayor (Ald. Tatham) thought that the Corporation should have the sole right to tear up the highways, and it was agreed to recommend to the Council that they should keep an eye on the situation. It was therefore proposed in full Council that the matter be referred to both the Gas Committee and the Parliamentary Committee. This was an unusual step, as the Parliamentary Committee normally dealt with all new Bills alone and so brought forth great indignation from Mr. Emsley, the Chairman of the Parliamentary Committee. He said that 'if the Council wished to have a Parliamentary Committee at all, they should place some confidence in it or abolish it at once.' To keep the peace, it was decided to let the matter be dealt with by the Parliamentary Committee with the addition of Ald. Bower and Mr. Woodhouse (Chairman and a member of the Gas Committee). The Parliamentary Committee, in its turn, appointed a sub-committee of three Aldermen and three Councillors (including Ald. Bower) to watch the Bill and to take all necessary steps for securing the interests of the Borough. At the same meeting, a petition against the new Bill was also decided upon for approval of the full Council and subsequent conveyance to Parliament in order to establish a *locus standi* when the Bill should go before committee.[3] The petition was withdrawn at the quarterly meeting of the Council on 3 May when it was found that the Corporation already had a *locus standi*.[4]

Local authority fears

In Parliament, it was resolved that the Bill should go before a Select Committee and among the witnesses called by the Committee was the Leeds Town Clerk, who appeared on 11 May.[5] He said that Leeds was willing to have the electric light, but that it would not be desirable to hastily adopt any system of electric light now before the public, and that it would be better to wait. He would strongly object to any company going to Leeds and supplying light, and, he added, Leeds wanted an absolute right to prevent unwanted companies from setting up. As an example of the Corporation's good faith to the citizens he quoted the price of gas, at 1/10d the cheapest in the country and claimed that this entitled the Corporation to a preferential right with respect to any fresh light. In conclusion he stated that the Bill as drawn was clearly inadequate to protect such Corporations as Leeds.

Many local authorities had similar fears. They were particularly worried that private companies might be allowed into their areas with rights not welcome to – or even held by – the municipal authorities themselves, such as the right to erect overhead wires or to dig up streets, and that they would find it difficult to protect the interests of the ratepayers. There was a fear, too, that success of private electric lighting companies could only be at the expense of the gas undertakings, so that the local authorities who owned their own gas works, such as Leeds and Manchester, believed that they faced the prospect of private profit resulting in public loss.

The fears of the local authorities were so great that the Council of the Association of Municipal Corporations met at Westminster on 20 April (Leeds was represented by the Town Clerk and the Chairman of the Parliamentary Committee) when it was resolved to memorialise the President of the Board of Trade to adopt four amendments to his Bill, which if accepted would give some protection to the rights of the local authorities of their gas interest.[6]

The Bill becomes law

On 18 August the Bill became law – the Electric Lighting Act 1882. Although there had been much discussion and subsequent amendment of the Bill during its journey through Parliament (and some amendments did indeed please the local authorities) the main provisions of the Act remained the establishment of 'Electricity Undertakers' – either local authorities, companies or private persons – in one of two ways[7]:

1. By the grant of a Board of Trade Licence for seven years, renewable, and subject to approval by the relevant local authority.
2. By the issue of a Provisional Order by the Board of Trade, subject to confirmation of Parliament. Local authorities could not prevent the issue of a Provisional Order but they had the right to purchase the undertaking at the end of twenty-one years, or every seven years thereafter, at the 'then value', but without an allowance for 'goodwill'.

With the passing of the Act it was expected that there would be a sudden increase in the installation of the electric light, but strangely this was not so. There was instead a shrinking of the industry, until there were allegations that the new Act had killed off the very industry it had been designed to protect and promote.[8]

There was no immediate indication of this in the summer of 1882, however, when the prospect of a spread of the electric light seized the public imagination to such an extent that there was a period of financial speculation of almost unprecendented intensity. It was a period 'remarkable for the rush of speculators in electric light shares', a 'speculative epidemic bearing some resemblance to the railway mania of forty years ago.'[9]

Fraud and failures

Unfortunately, electrical equipment was not as well developed as Parliament had believed it to be when the Board of Trade had been persuaded to present the new Bill. Technology was not yet sufficiently advanced to allow successful widespread central supply and most of the public's money was lost on inefficient equipment, dubious patents and worthless concessions. While a few of the holder of patents and concessions made fortunes, the public only learned by bitter experience that 'electric manufacturing follows much the same commercial laws as other branches of engineering'.[10]

It is interesting at this point to speculate on the financial affairs of Robert Hammond, who although without doubt was a genuine supporter of the electric light and who spent a lifetime encouraging its use and utilisation, nevertheless must have found himself in the same financial position as many of the frauds and tricksters. He had paid £20,500 for the Brush concessions for ten English counties. The Hammond Electric Light and Power Supply Co. had then sold the Yorkshire concessions to the Yorkshire Brush Co. for £50,000, plus fully-paid-up shares worth £50,000. If concessions for all ten areas had been sold on the same basis Hammond's would have received £500,000 plus shares to the same value; without undertaking any electrical work at all the Hammond companies would have profited by the sale of concessions to the sum of £1 million. Robert Hammond, by virtue of his founders' shares (and their entitlement to a third of the profits of the company) could well have received nearly £170,000 in cash and shares to the same value. And all this for an initial investment of £20,500!

Although it's difficult to accuse Robert Hammond of deliberate company mongering, it is clear that much fraudulent behaviour was going on and that the business of electric lighting was 'disgraced by the commission of every crime that commercial immorality could invent.'[11] It's not hard to understand why the public eventually withdrew their financial support from electricity companies, so that even the reputable companies suffered a period of considerable restriction.

The companies were also hampered by the ironic fact that the very improvements in equipment which had stimulated enough interest to promote the new Act at the same time discouraged the use of the new equipment because prospective customers and purchasers realised that by waiting a short time better and cheaper equipment would be available. So serious were these restrictions against the business of the companies that in 1883 of the fifty-five company Provisional Orders granted, none were used. The number of new orders in 1884 fell sharply to four, and apart from a single order in 1886, there were no more until 1888.[12]

Local authority dilemma

This lack of success by the companies put local authorities such as Leeds in rather a strange and confusing position. They were obviously pleased at the companies failures, for they were determined to keep them out of their areas at all costs, and yet the inability of the companies to provide a public supply increased considerably the pressure on the councils to step in and install centralised electricity supplies themselves. This was an unpleasant prospect for some councils who had no wish to invest hastily in unproven and probably obsolescent equipment, only to be condemned for negligent expenditure of ratepayers money. Yet they had no wish to hold back unreasonably from electricity supply, for they could then be accused, with some justification, of ignoring the wishes of the ratepayers. There was the risk, too, that by holding back they would encourage private companies to step in, and this was a possibility not to be contemplated.

There was a middle way, of course, and that was to undertake a small-scale experimental installation of the electric light before deciding on a definite policy. Such a course had several advantages: the council would gain experience of costs and equipment at small expense; the experiment would take some time to complete and would thus allow further development of electrical equipment to take place; and the public, the Board of Trade and private companies would be convinced of the serious intent of the council. Any council able to undertake such an experiment would surely have a greater chance of making a successful decision about the electric light.

Notes and references

1. *Leeds Mercury*, 6 May 1882.
2. *Ibid*, 14 April 1882.
3. *Ibid*, 25 April 1882.
4. *Ibid*, 4 May 1882.
5. *Yorkshire Post*, 12 May 1882.
6. *Leeds Mercury*, 21 April 1882.
7. *The Organisation of Electricity Supply in Great Britain*, H. H. Ballin, (1946), p. 10.
8. W. H. Preece, In *Electrician*, 15 May 1884.
9. Jackdaw, 20 May 1882.
10. *The Times*, 22 January 1883 'Wholesale Electricity'.
11. W. H. Preece, in *Electrician*, 15 May 1884.
12. *The Organisation of Electricity Supply in Great Britain*, p. 11.

Chapter 6

The Electric Light Commitee, Leeds

The improvement in electric lighting apparatus in 1881, which led eventually to the Electric Lighting Act of 1882, also coincided with the erection of Leeds' new Municipal Buildings, grandiose structures which were to house not only Council offices but also the Free Lending and Reference libraries. The site was in Calverley Street, opposite the magnificent Town Hall, and was part of the City Fathers' dream of building a civic centre to rival that of any other town in the country; it was the ideal opportunity for the Council to install their first small-scale electric light installation.

At first there was no sign of the difficulties to come, Council members agreeing that electric lighting should be considered for the new buildings and so instructing the architect, Mr. Carson, to investigate modes of electric lighting and report on their suitability for the new premises. His report was presented on 15 July 1881 to a sub-committee of the Property Committee and the Free Public Library Committee,[1] when he explained that he had examined the House of Commons, Liverpool Free Library and Picture Room, and several other places, and he could report favourably on the electric light. He thought the working cost would be about the same as gas but a better, purer, light would be obtained. There was no need to take immediate action, he told the sub-committee, until the inside was about to be plastered, when tenders could be obtained from various firms not only for the new offices, but also for the adjacent Town Hall and nearby streets. In the meantime, he concluded, the delay would allow further improvements in electric lighting to take place.

The sub-committee unanimously approved the report which gave the Chairman of the Library Committee great satisfaction. He stated that the whole of the Library Committee were wholeheartedly in favour of the electric light and that they would be happy to allow the Corporate Property Committee to deal with the matter.

The Property Committee's report

The Property Committee took their responsibilities very seriously, deciding that they should prepare their own report on the electric light, despite Mr. Carson's report. After all, it was worth taking trouble on a building costing £100,000. A joint deputation was therefore formed with representatives from the Property Committee, the Free Library Committee and the Gas Committee, charged with the task of discovering and reporting on the recent advances in electrical equipment and advising on its suitability for the new offices. The deputation consisted of the Mayor – Mr. George Tatham – the chairman of each of the three committees, plus five other members from the committees, and also the Deputy Town Clerk and the Gas Engineer.

Seven months after Mr. Carson's report, the deputation set off for London where on the 22, 23 and 24 February 1882 they visited the Crystal Palace Electrical Exhibition, examined the electric light installation at the Savoy Theatre and noted its effects, and inspected various public buildings (such as railway stations) where the electric light had recently been installed. The next month or so was spent in studying documentary evidence, such as newspaper and magazine reports, and the pamphlets issued by the various electrical companies.

Long before the report was ready for publication, the Mayor could not resist 'leaking' the conclusions to a Council meeting on 31 March.[2] It was his view that 'so rapid was the improvement being made both in electric lighting and gas lighting that what was affected today might be superseded in a short time.' This view obviously indicated a postponement of any experiment with the electric light, a postponement that was confirmed when the full report[3] was issued to the Council a month later, on 3 May.[4]

It was a report which the Mayor designated as 'a very comprehensive and exhausting report', which had been compiled largely by the Deputy Town Clerk, Mr. Jollicliffe. It began by illustrating how electricity was generated and what equipment was required; it continued by describing the kind of equipment available and gave an assessment of its worth, then went on to detail actual installations, supplying where possible the particulars of costs. The report also published the deputation's considerations of electricity generation, and the two ways which it thought were particularly relevant to Leeds.

The first method was by water power. The deputation stated that the Corporation still owned water power and vacant land formerly used in connection with the old 'Pitfall Mill' at Leeds Bridge. The water wheel, it was reported, had a diameter of 19 feet 6 inches, a face width of 5 feet 9 inches and 2 foot floats. It was estimated to have a power output of 14 h.p.

The second method of generation was explained by quoting an article from the *Engineer*: 'Town refuse treating machinery and apparatus made by Messrs. Manlove, Alliott and Co. is being very extensively and successfully adopted, one reason for its success being that the cinders, cabbage stalks etc. in the refuse furnishes the fuel for doing the work. What is more – and this is dreadful – is that the fuel so provided is much more than is required for the work, and it is proposed to light some places by electricity by means of this waste.' It was noted that this refuse burning system was already installed at Burmantofts and Armley, waste heat therefore was now available in Leeds for the possible generation of electricity.

Having noted the two possible sources of electricity generation, the report made no suggestion of the utilisation of either. In fact, no suggestion of generation was made at all, the deputation merely noting that arc lamps were well adapted for the illumination of large areas, inside or out, and that they had been much impressed with the success of the incandescent lamp. Portions of the new Municipal Buildings could be successfully lighted with this class of lamp. The conclusion was very much as the Mayor had previously suggested: 'The great advance, notably during 1881, in the development and improvement of the numerous gas and electric lighting systems may be taken as an augury that further discoveries will be made. Therefore the deputation are of the opinion that it would not be desirable hastily to adopt for a permanency any system of electric lighting now claiming public attention.'

The only actual improvement to lighting in Leeds resulting from the deputation's enquiries was the installation of two of the new Siemens gas lamps in Briggate. Although much brighter than previous gas lamps, it was noted in the *Yorkshire Post* of 4 May that they consumed about seven times as much gas as previous lights.

The deputation's costs

A most amusing result of the deputation's enquiries was a notice on the agenda of the Leeds Council Meeting of 3 May from Mr. Barker, the motion being 'that return be presented to the Council of all deputations appointed by the Council or any of the committees from 9 November 1879 to 30 April 1882, setting forth by whom appointed, the names of the deputation, together with the amount paid to each member.' Mr. Barker was obviously working hand-in-glove with Jackdaw, who had been complaining bitterly of the outrageous price of the jaunts to the country or to London by various deputations. When the motion was actually discussed on 18 May[5] Mr. Barker said innocently that he had learnt from an outside source that a recent deputation to a quarry at Threlkeld in Cumberland had cost the Corporation £400. Amidst considerable Council amusement, the Mayor told him that it was not always wise to believe everything in newspapers, not even, he added scornfully, the *Leeds Mercury*. He added that the deputation had actually cost £32, not £400. Despite laughter and attempts to shout him down, Mr. Barker stuck to his guns, claiming that allegations had been made that the deputation to London had charged first-class fares when they had only paid third-class. 'Was that honest?' he asked, amid increasing uproar. The Council were having none of it and the resolution was lost. On the following Saturday, Jackdaw, his feathers obviously ruffled, commented on the 'silly and petulant observations which one or two Town Councillors were pleased to make with regard to my part in this inquiry.'

This incident may have been unremarkable had it not been for the fact that the redoubtable Mr. Barker at last – on Thursday 28 September 1882 – managed to get the Council to agree to his slightly amended resolution,[6] so that details of deputations from 9 November 1879 to 'the present date' were presented to the Council by the Town Clerk on 9 November.[7] The return was laid upon the table and no comment was made, a silence which was surprising for although the London deputation was shown to have cost a modest £59.3s.2d, the Highways deputation visit to Threlkeld Quarry had cost not the £32 mentioned by the Mayor nor the £400 alleged by Mr. Barker, but a staggering £529.9s.6d.[8]

During the months of Mr. Barker's struggle with deputations' costs, the actual erection of the Municipal Buildings was completed, in July 1882, and an argument immediately broke out between the Town Clerk's department, the Accountant and the Library Committee. The Town Clerk discovered he did not have enough room, and insisted on taking some of the Accountant's space. The Accountant, in his turn, refused to give up space unless he could take the equivalent from the Library, which he said, in any case had an undue proportion of the 86,000 sq ft available with an allocation of 45,000 sq ft. The Library Committee Chairman remarked that much of his space was made up of stairways and corridors, and was not much use to him, but if he had no option but to reduce his area he could do no other but give away the Ladies Reading Room, a thought which provoked much public horror and discussion. It was eventually decided, though, that women would have to read in the same room as the men, the space would be given to the Accountant, who would transfer equivalent space to the Town Clerk and so agreement was reached.[9]

The ELC takes responsibility

The only problems now remaining were organising the furnishing and fitting out of the buildings, once the interior building and decorating was complete (although this was expected to take several months) and deciding on the form of lighting which would be installed. The Purchase of Property Committee was made responsible for both these duties, but within a month, another Committee was to take over the investigations into the electric lighting.

The main fabric of the Municipal Buildings was completed by July 1882. The next month, on 18 August, the Electric Lighting Act became law. The following week there was a special meeting of Leeds Town Council, called to consider the provisions of the new Act, at which the Council heard from the Town Clerk that the provisions 'have constituted such Corporations as that of Leeds practically masters of the situation within their area.' The Town Clerk also reported that the Hammond Company had applied for a licence to supply electric lighting in the Borough. When it was pointed out that not only was there no committee empowered to deal with such applications, but there was no committee to deal with any questions arising under the Electric Lighting Act, Mr. Emsley, Chairman of the Parliamentary Committee, proposed that such a committee be formed, a resolution accepted by the Council who elected a committee of twenty-three, including the Mayor (George Tatham) and Mr. Emsley.[10]

This new committee was called the Electric Light Committee (ELC) which Mr. Emsley believed would in a few years be one of the most important they had in the Council. Furthermore, he suggested that although a company had already applied for a licence, 'it appears to be one of the most important matters to consider whether the Council should not take up the subject themselves.' These thoughts were well received by Council members, as were similar sentiments expressed a month later at a dinner held in the Mayor's Rooms, in the Town Hall.[11] The Mayor entertained members of the Town Council at the dinner which celebrated the approaching close of his third year in office, and he invited chairmen of some of the principal Council Departments to make a few remarks. Mr. Emsley, speaking for the Parliamentary Committee, said that the Electric Lighting Act had been the most important subject of last session. His own impression was that the electric light would become part of the lighting of great towns, including the town of Leeds, and that Leeds would be able to take the lead.

The Mayor's dinner party was held on 19 September 1882, the same day that the first meeting of the ELC took place. The Mayor, of course, was elected Chairman of this new committee.[12]

The ELC had two immediate problems before it, the first of which was solved by recommending that the leading electrical companies be asked to tender for the installation of the electric light in the Municipal Buildings and inside and in front of the Town Hall.[13] The second problem was more complex. During the summer months many more electricity supply companies had been formed, encouraged by the fact that there was a growing demand for electric light and that Corporations were loath to spend their own money on it. The Electric Lighting Act seemed to give these companies respectability too, and they lost no time in applying for Provisional Orders or Licences once the Act became law. By September there were more than thirty such companies[14] and six had given notice to Leeds Corporation that they would be applying for authority from the Board of Trade, the six being Hammonds (previously noted as having applied for a Licence), Gulcher, Edison, Crompton-Winfield Associates, Swan, and an annonymous applicant represented by Messrs. Walter Webb, Solicitors.[15] The ELC had to decide which of these if any, should be given approval, or whether Leeds Corporation should take on the role of electricity undertaker. Mr. Emsley's view that 'the Council should take up the subject themselves,' was obviously shared by the rest of the ELC, which decided to recommend that the Council should apply for a Provisional Order, the local authority of course having preference.[16]

Nine days later, at the Council meeting of 28 September (the same meeting at which Mr. Barker's patiently presented resolution had at last been accepted) the ELC's recommendations were put forward. The electric light proposal for the Municipal Buildings was accepted (the tender advertisement to be published nation-wide), but the second recommendation placed the Council in rather an awkward position. They were under pressure from private companies to allow private control of public electricity supply (indeed a further two companies had added their names to the list)[17] although private control was something the Council disliked. And yet they didn't feel they should spend ratepayers' money – and possibly waste it – on a system which would soon be old-fashioned, development at that time being so rapid.[18]

The Mayor agreed that electricity was the light of the future, but noting the recent improvements and reduction in cost, he felt it would be unsafe to proceed at once and take a leap in the dark. He also felt they should not be behind the times, but should take advantage of the best mode in lighting. The Council should have the entire control of the supply of electricity within the Borough. Although the rest of the Council members shared the Mayor's doubts, it was this last desire for control which outweighed any other thought, and it was reluctantly agreed to hold the necessary statutory special meeting in one month's time to authorise the application to the Board of Trade for a Provisional Order, the ELC in the meantime to prepare the draft application. Having thus tentatively agreed to proceed with the electric light, the hope was expressed that it would be available in time for the Music Festival next year.

Provisional Order problems

As expected, at the statutory meeting in November 1882 the Council confirmed the decision to apply for a Provisional Order and to turn down the Licence applications from the private companies,[19] an uneasy decision no doubt forced on them by pressure from the private companies and increasing pressure from the newspapers. A bitter editorial in the *Yorkshire Post* on 11 October started as follows: 'Light! More Light! These are said to have been the last words of a German poet. But if it had been the lot of this poet to live in Leeds they would in all probability have been his first words.' The editorial went on to say that Leeds was one of the worst lighted towns in the British Isles: 'The gas lamps are of the most antiquated description; the gas itself, if not impure, is of inferior quality and the force is apparently so poor that the lamps, which are perhaps sufficiently numerous to bathe the streets in a flood of light, only serve as a matter of fact to make darkness visible.'

The probable explanation of the Council's extreme reluctance to spend the ratepayers' money was provided in the *Yorkshire Post* three days later (14 October): 'Leeds is encumbered with one of the largest public debts owing by any British municipality; its rates are amongst the very highest.' The debt, of £4,000,000,[20] was large enough to daunt the most adventurous councillor, and was undoubtedly the reason that Leeds Council found itself in its embarassing position; they had applied for a Provisional Order they didn't really want, or thought they could afford, merely to stop private companies undertaking electricity supply in the Borough. In an attempt to resolve the problem the Council wrote to the Board of Trade asking that, in view of their *bona fide* desire to provide their citizens with the best and most economical light (as evidenced before the Select Committee), the Provisional Order should be permissive, but not obligatory.

The Board of Trade replied that a Provisional Order was for a definite undertaking, not a general authorisation; the schemes were not intended to merely enable a promoter to appropriate an area of supply to the exclusion of other applicants.[21]

Thus discouraged, the Council had no alternative but to join the other forty to fifty towns, including Liverpool, Manchester, Glasgow and Bradford, who sent a deputation to meet Mr. Chamberlain, President of the Board of Trade at the beginning of December.[22]

Mr. Littler, QC, spoke on behalf of the deputation. He said that corporations should have the right of supply of electricity in their areas, as with water and gas. The corporations felt that electrical equipment would be greatly improved over the next few years and did not wish to waste the ratepayers' money on systems which could soon be outdated. He presented clauses to amend the Bill, one of which enabled a corporation, under a Provisional Order, to decide the appropriate time to supply electricity. The amendment provided for an inquiry to investigate whether the corporation was at fault if one twentieth of householders in the Borough required it.

Mr Chamberlain was unimpressed. His opinion was that if a private company wished to hazard their money on supplying electricity they should be allowed to do so if the corporation would not. Not to do so would be unfair to those householders who wished it, even if they were of a minority much less than one-twentieth.

While all this negotiation was taking place, replies were coming in in response to the tender invitations by the Corporation. On 25 January 1883 the ELC met to consider the replies and decided they were not really competent to decide on the relative merits of the various systems. They needed further information on the different systems and decided that yet another deputation should be formed to carry out this investigation. Accordingly, an eight man sub-committee was formed, including the Mayor, which in the next couple of months visited various premises in London, Birmingham and Chelmsford, where in particular they saw the works of Winfield and Crompton.[23]

The report of this second deputation was presented to the ELC on 16 April, the conclusions about arc and incandescent lamps being similar to the previous deputation's report. The Burgin Dynamo from Cromptons was recommended as the most suitable for the Leeds lighting which it was proposed would be restricted to the Victoria Hall, General Pay Office, Clerk's Room, Lending and Reference Libraries and the Reading Rooms, with three arc lamps in front of the Town Hall and two in front of the Municipal Buildings. The shed in the new Fire Brigade Station appeared to be well adapted for the engine and dynamos. The final recommendation was that the Council be asked to authorise the ELC to ask (again) for tenders, then report to the Council.

The ELC accepted all the recommendations of the deputation except for the use of the Fire Brigade hut, which they felt would materially interfere with the necessary requirements of the Brigade. This was referred back to the sub-committee for reconsideration.

The next week the ELC met again to discuss the draft Provisional Order which had been recently received. It must be admitted that the ELC didn't like the Provisional Order, objecting strongly to being compelled to give a supply to anyone.

They did not wish to accept an Order which placed this obligation on them until electric lighting was practicable, but despite this Aldermen Tatham and Gaunt agreed to promote the Provisional Order to the Council, which was meeting next day.[24]

The draft was explained to the Council by the Town Clerk, who told them that the important clauses required the selection of two areas, Schedules A and B. A was that area where immediate supply would be given, to nominated streets in that area. Parts of B could be added under certain circumstances. After 2 years, mains could be laid in the rest of A, or in adjacent parts of B on requisition from householders, who must agree to take such a supply for three years as to produce in the aggregate an annual sum of 20 per cent upon the expense of giving a supply. No immediate decision was taken by the Council, partly because of the lateness of the hour and partly because they wished the ELC to define the areas A and B.

At a new Council Meeting on 2 May, the ELC had considered the Schedules, and area A was defined as that area bounded by Great George Street, Oxford Place, Park Lane, East Parade, Bond Street, Park Row, Park Lane and Calverley Street.[25] The streets to be supplied were Victoria Square (in front of the Town Hall) and Calverley Street, while the places proposed for supply were the Town Hall and the Municipal Buildings. Schedule B was the whole of the Borough not including Schedule A.

Having clarified the Schedule Area, Ald. Tatham then moved that the Provisional Order be accepted by the Council, supported by Ald. Gaunt. They argued that the Order would allow the Council to light the public buildings, as they wished, and would prevent other companies from supplying electricity in the areas for at least two years. There was concern though that nowhere had electric light been successful, although the Mayor reported that the Mayor of Chesterfield thought their electric light had been a success, and Mr. Hammond told them his house was lit cheaper by electric light than by gas, (but his gas was 3/2d per 1000 against Leeds gas at 1/10d). It was further pointed out that if the initial cost of installing the electrical apparatus to comply with the Order was £10,000, the further cost after 2 years (whether equipment was sufficiently improved or not) would be £30,000 or £40,000 to extend supply to all of Schedule A.

In view of the previous statements from the ELC it is not surprising that the Council threw out the Provisional Order by 25 votes to 4: even Aldermen Tatham and Gaunt voted against it.

Ald. Tatham then proposed that the Council authorise the ELC to carry out the experimental system of electrical lighting within and about such portions of the Municipal Buildings as the Committee may consider desirable, at a cost not exceeding £10,000. There was a feeling that it would be unwise even to spend £10,000 on electric light and that the new buildings should be lighted by gas. However, Ald. Tatham reminded the Council that they had previously sanctioned the lighting of parts of the public offices by electricity and some trial should surely be made. The second deputation's report, which had been circulated round the Council, was no doubt influential in helping the Council to decide to support

Ald. Tatham and the ELC. They were allowed to go ahead with the electric light, up to a limit of £10,000.

As a result of this decision the ELC once again published particulars of their requirements for lighting the Municipal Buildings, asking the principal companies to tender for the work.[26] Electric light was required, the particulars stated, in the Vestibule and Victoria Hall of the Town Hall, and the Vestibule, General Pay Office, Gas and Water Clerk's Office, General Reading Room of the Library, Lending Library, and both lower and upper levels of the Reference Library in the new Municipal Buildings. Outside, lights were required in Victoria Square and Calverley Street. A summary of the expected lamps to be used was:

20 CP Incandescent Lamps	
Victoria Hall	500
General Reading Room	51
Lending Library	70
Reference Library	140
Total	761
Arc Lamps	
Town Hall Vestibule	1
Municipal Building Vestibule	1
General Pay Office	2
Clerks Office	2
In front of Town Hall	4
In Calverley Street	1
Total	11

Although lamp numbers and positions were shown on the plans, the ELC pointed out that this information was given with a view to the tenders being prepared on a common basis, and that if any improvement was possible this should be included in a second tender price. The Corporation were to provide the engines which would drive the dynamos (unless the company could better provide the whole installation) and a power house would be available no more than 400 yards from the Town Hall. Excavation in the public highway, and provision of three inch iron pipes, would be by the Corporation.

The tenders were to be for a period of either one or two years, and were to give costs for either complete removal of equipment at end of contract, or purchase by the Corporation. Work was to be completed two months from acceptance of the tender and tenders were to be in by 21 June.

No sooner had these details been published than the ELC sent out supplemental particulars, on 29 May,[27] which made the following alterations:

1. There was now no insistence of using Crompton–Burgin dynamos.

2. The Corporation would cover wires in the buildings and make good at their own expense.
3. The ELC would not insist on any particular make of incandescent lamp (they had previously reserved the right to so insist).

Tender success too late to save Yorkshire Brush

By the time that these supplemental particulars were issued, the ELC had been in existence for just over nine months, from the beginning of August 1882. It was a period which had started with great optimism and high ideals. This optimism was changing to pessimism, however, as the initial enthusiasm was being dampened by the delays which the ELC were causing (tenders had twice been called for). These delays were having a serious effect on the Yorkshire Brush Co., troubled as it already was by its disputes between shareholders and management about working capital, and the problems with the Lane–Fox lamps. The company was now desperate to obtain a contract for a large job in which a central station could be used to supply all types of customers. A contract for £11,000 had been put before the Leeds authorities when the first tenders had been called for. Mr. Hammond felt

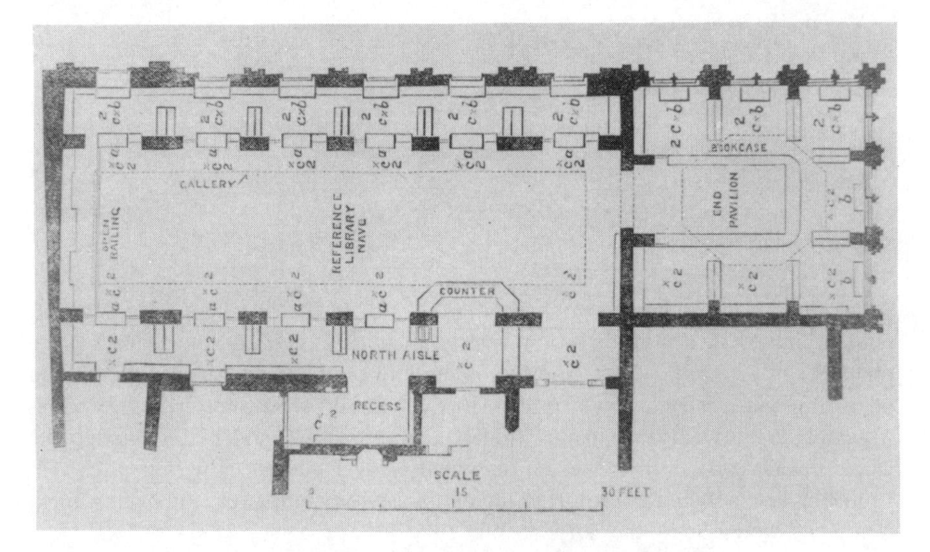

Fig. 14. *Plan of library to be lighted by electricity (*Electrician*, 26 May 1883)*

this was an offer which could not be refused and which was bound to lead to greater things.[28] In his opinion, even the Chairman of the Gas Committee was convinced of the advantages of electric light and even the £1,000,000 invested in Gas would not deter the Corporation from putting down the electric light system on a large scale.[29] It was a great blow, then, when the ELC decided not to go ahead with the first tender applications. Yorkshire Brush staff had taken much time and trouble in the preparation of estimates (which was all work for nothing), and a great public relations effort had been made, with visits to Leeds by both Ferranti and Ince. (These visits, in fact, belatedly culminated, in September 1883, in a display of the Ferranti system of electric lighting at the premises of Marsh Jones and Cribb of 48 Boar Lane.)[30]

Yorkshire Brush believed that the majority of the Council preferred this local Leeds company, but could the ELC make up its mind? By the time the ELC did make the decision to accept the low tender, it was too late; Yorkshire Brush, in September 1883, was not able to accept the contract. By October, its £2 paid-up shares had fallen to 5/–[31] and at the Annual Meeting of the company that month it was reluctantly proposed that the company amalgamate with Hammonds, who owned 17,060 fully-paid-up shares. An extraordinary meeting of Yorkshire Brush agreed to this proposal,[32] as did an extraordinary meeting of Hammonds in November[33] where the terms of one £5 Hammond share for six £2 Yorkshire Brush shares were confirmed, and Yorkshire Brush was formally wound up at the beginning of December 1883.[34]

The Hammond Co. fared no better than Yorkshire Brush. Despite an increase of capital in April 1884 from £330,000 to £525,000,[35] the company was voluntarily wound up in June 1885.[36] (Robert Hammond survived as an electrical consultant and re-appears later in the Leeds story.) The parent Anglo-American Brush Co. also had problems, losing £1,600 on trading in 1883[37] and halving capital in 1884 (so that £10 of shares were only worth £5).[38] 1884 turned out to be a year of profit, though (£4,000),[39] and business improved steadily in 1885 and onwards.[40]

Lights for the Music Festival

The liquidation of Yorkshire Brush meant that the ELC did not have a contractor for the scheme to light the Municipal Buildings. Their first reaction was to decide to go no further, a course of action which when publicly announced caused an uproar, particularly from the Mayor and the Committee of the Music Festival, who had been promised electric light in the Town Hall for the Music Festival in October. Indeed, it was no secret that if they had known the ELC were not to supply the electric light, they would have made arrangements themselves.[41] Jackdaw told his readers that the ELC had 'bungled it in a very unpleasant manner. It is distinctly discreditable to Leeds and its local rulers that there should have been any hitch whatever in the plan for giving the patrons of the Festival, and the members of the band and chorus, the benefit of electric light.'[42]

This accumulated pressure on the ELC eventually prompted them to action. Under the direction of the Mayor – now Mr. Edwin Woodhouse – they approached the Crompton–Winfield Electric Lighting Association (who twelve months previously had successfully lighted the Birmingham Town Hall for their Music Festival of August) and agreed the lighting of the Town Hall as appropriate.[43] Mr.

Greig of Fowlers promised to supply steam engines free of charge,[44] and he and the Mayor worked to smooth any difficulties, the ELC having very little to do with any of the arrangements.[45] Although the Music Festival was to start in three weeks time, on 10 October, Cromptons stated that they could finish in time, and started immediately.

There would probably have been less agitation about the electric light had the Music Festival been of less importance. However, the Leeds Triennial Festivals, which had started in 1874,[46] had attracted increasing support and attention, both locally and nationally: the Committee knew that all England's eyes would be on Leeds. Moreover Prince Leopold, Duke of Albany ('England's cultured and student son' as *Punch* called him),[47] Queen Victoria's eighth child, had agreed to be President of the Festival. Although his visit to Leeds was to be a private one (the town therefore not being decorated),[48] Leeds was proud that after his visit it would have had visits from all the Queen's sons, as well as the Queen herself.[49]

It was with some relief then that the Committee learned that all would be ready on time. Crompton Winfield had worked well, under the superintendence of Henry Lea of Birmingham (Consulting Engineer) assisted by Wilson Hartnell, the Leeds Consulting Engineer and Mr. Kyle of Cromptons.[50] A description of the installation in the huge Victoria Hall, 140 feet long, 65 feet wide and 75 feet high, is best left to the *Electrician* of 20 October 1883:

> 'The plan adopted and probably the only one suited to the structural conditions, was to suspend seven pendants from the arched ribs of the roof, three in the middle line, and four towards the corners, so as to ensure an equal distribution of light as well as an effective appearance, this latter being, of course, an important consideration.
>
> These pendents illustrate the artistic capacity of the incandescent light, fitted up at the utmost pitch of speed. Elaborate workmanship was out of the question. Nothing is used in them but plain hoops and simple brass tubing, yet they are as effective as if built up of the most elaborate metal work. The truth is evident that the dazzling points of light, when canopied, shaded and enlarged by the tinted glass covers, are sufficient in themselves; a tasteful arrangement of their lines and groupings is all that is necessary to satisfy the eye. This is obtained by rings of lamps rising in a cone from a single lamp, each ring varying in distance from the others, as do the lamps in each ring. There are fifty-six lamps in each pendant arranged two in series, and connected together by wires on the tubes, led up to a central plate at the top. They are suspended by steel wires passing through holes in the arched ribs of the ceiling in the centre of ornamental points, so that they can be lowered whenever required, connection being made through loose cables hanging by the side of the supporting wires. The effect is as though the lights were self-supported.
>
> The lamps themselves are of the latest Swan pattern, of 160 ohm hot resistance, and working with a current of about ampere 0.65, giving twenty candle power.
>
> The engine and dynamo machines are stationed in a street about 300 yards

distant, and the circuit between them and the hall is made by a number of cables of seven No. 16 wires, all inserted in one iron pipe provided by the gas department.

Under the exigencies of the case it was necessary to employ such machines as happened to be in readiness, and it was found better to use independent machines and circuits for each pendant, and for a row of fifty-six lamps under the gallery, which make up eight distinct sets. There are therefore nine dynamos provided, of the C pattern of the well known Crompton–Burgin machine, as nearly alike as the circumstances allowed; but at this stage they all blend into one driving system, and are connected separately to a commutating system by which any dynamo can be used to either of the circuits and their working condition tested at any instant. They work with an average e.m.f. of 240 volts, and give an average current of 17.5 amperes to each of the eight circuits.'

The dynamos, mounted side by side upon a strong wooden platform were driven by leather belts from a countershaft of 4 inch diameter, running at 370 rpm, in turn driven by a 200 h.p. compound steam engine, provided and run (as promised) by John Fowler and Co. of Leeds as their personal contribution to the festival. The massive size of the engine was due partly to the need to secure the wide margin of force over work essential to the quiet, steady exertion of power (to produce steady light) and partly to the fact that arc lamps were to be added to the system later on for another meeting.

'Slight hitch' and then success

This was the system, then, awaiting its trial at the opening of the festival by the Duke and Duchess of Albany. The royal couple were stopping in Otley, a few miles outside Leeds, as the guests of Mr. Ayscough Fawkes at Farnley Hall (famed for its association with Turner). Their arrival in this delightful little town was marked with a day of public holiday, enjoyed by the thousands from surrounding towns who were filling Otley to overflowing.[51] The morning of Wednesday, 10 October was fine and sunny in Otley as the Prince and his wife boarded the train for Leeds, but as Kirkstall was reached on the outskirts of Leeds, the sun disappeared and the weather became worse as Leeds was approached. Rain fell in torrents at first, then in a miserable drizzle, and by noon the centre of the city was enveloped in a fog described by the newspapers as 'black and horrible in its darkness as it ever is in the depth of winter . . . the gloom was very gruesome and forbidding.'[52] The darkness was so intense that the street lamps had to be lit. Inside the Town Hall there was the prospect of the first morning performance being conducted by an orchestra invisible to the audience, and although the contractors had only undertaken to provide the electric light at seven o'clock that evening, there was no alternative but to switch on straight away. While the engine room was still being cleaned and whitewashed, and the dynamos being tested, it was decided to get things into order and light up the hall eight hours before the due time.[53]

The result was disastrous! The incandescent bulbs, though at first bright, slowly became dimmer and dimmer until they were almost out, and eventually the gas lights had to be switched on. Mr. Hartnell explained:

> 'No one had anticipated that the electric light would be required at midday, especially in October. The request to light up was as sudden as it was unexpected, and no preparation was made for a continuous run. A short stoppage was therefore desirable the first opportunity (which happened to be in the interval between Parts 1 and 2 of Elijah, by Mendelssohn, the opening work). The installation as a whole had had no previous trial, and it was intended to have run it previously only yesterday afternoon.'[54]

The *Electrician*[55] tells us that the cause of the mischief was a bearing which was 'becoming undesirably warm, no doubt owing to a little grit which had got into it during the hurry of unexpected cleaning up.' This magazine treated the occurrence as but a 'slight hitch' and Jackdaw too thought it only a 'slight failure', and understandable at that. 'I need hardly say that imperfections are constantly discovered in the trials of new machinery.'[56]

Once the fault had been put right, it was generally agreed that the electric light had proved a great success. Jackdaw was very pleased. 'I visited the shed in which the apparatus which supplied these beautiful lights was erected. The splendid engine was working almost noiselessly and the wheels of the dynamo were revolving at such a rate of speed that to the eye they seemed to be absolutely motionless. What a contrast the silence and the absence of evil odours presented to the ordinary gasworks! How long will it be before Leeds is fully lighted by electricity?'

The *Yorkshire Post*, too, liked the electric light. 'It will prove a vast improvement over the old method of illuminating the hall, both as regards the intensity of the light, and the absence of heat and smell.'[57] An 'Unmusical Man', writing in the *Leeds Mercury*, remarked that 'everybody looks up at these electric sparks with a murmur of admiration.'[58]

Despite the initial 'slight hitch' with the electric light, it was apparent by the end of the week that the Festival had been a huge success. More people than ever before (13,984) had attended the concerts, the Festival had produced a profit (for charity) of £2,000, and congratulations had come from all over the country. As the *Daily News* reported, 'the result is not only one for which Leeds may take great credit, but one in which the country at large has reason to share her feeling of pride.'[59] The Festival was also the occasion for one of the first outside broadcasts in England.[60]

The electric light installation in the Town Hall was left where it was when the Music Festival had finished, indeed it was added to by the addition of four arc lamps mounted in Victoria Square immediately outside the Town Hall. This was for a National Conference held by Liberals on Parliamentary Reform, on the Wednesday and Thursday following the Festival. The Liberals had asked to use the electric light, no doubt to add to the status of an undoubtedly prestigous meeting which was to be addressed in particular by the 'Peoples Tribune' – Mr. John Bright, probably the most popular and able orator of the Liberal Party.[61] The aim of the Conference was to press for an extension of franchise and for womens suffrage and was attended by delegates elected by the Country's Liberal Associations, but Thursday evening was thrown open to the public. So many people turned up to hear and see Mr. Bright that the Victoria Hall was full to overflowing, the overflow being accommodated outside in Victoria Square. The Saturday Supplement of the *Leeds Mercury* (20 October) reported that 'fully 20,000 faces were turned towards the (Town) Hall, which the strong rays of four electric lights made as clear almost as if it had been noon, and not 7 o'clock on a cold and an uncertain October evening.'

Jackdaw also approved, of course, stating:

> 'Leeds has now had a fair opportunity of judging the electric light on its merits. The trial which has been given to it during the past fortnight must be pronounced a great success. The Hall even last night, when it was crowded to excess, was comparatively cool, and the general effect of the lights was admirable. In Victoria Square, too, electricity won a great triumph. It is to be hoped that arrangements will at once be made for giving us permanently the advantage of this method of illumination.'[62]

To enable those of the public who had not yet been able to see it for themselves to do so, the Town Hall was opened up on the first three nights of the following week, and musical attractions were provided. Surprisingly, thousands of people turned up at the Victoria Hall to see the Swan lamps and there was universal satisfaction. The *Leeds Mercury* pointed out the next weekend that although expensive (compared to the cheapness of Leeds gas), the electric light had other benefits such as beauty, cleanliness and superior healthfulness.[63]

Notes and references

1. *Leeds Mercury*, 16 July 1881.
2. *Ibid*, 1 April 1882.
3. Leeds Corporation Report on Electric Lighting, by Joint Deputation, 1 May 1882.
4. *Yorkshire Post*, 4 May 1882.
5. *Leeds Mercury*, 19 May 1882.
6. *Yorkshire Post,* 29 September 1882.
7. *Leeds Mercury*, 10 November 1882.
8. *Yorkshire Post*, 11 November 1882.
9. *Ibid*, 15 and 22 July 1882.
10. *Leeds Mercury*, 26 August 1882.
11. *Ibid*, 20 September 1882.
12. *Ibid*.
13. *Yorkshire Post*, 13 October 1882.
14. *Leeds Mercury*, 20 September 1882 – thirty companies, capital over £6m.

15. *Ibid*, 29 September 1882.
16. *Ibid*, 20 September, 1882.
17. It is interesting to note here the two new companies who had given notice of application to the Board of Trade for a licence. The first was an arc lighting company called Jablochkoff which had originated in Russia. The second was a remarkable Leeds partnership of Messrs. Robinson and Mori. Mr. Robinson was the proprietor of a Grocery and Italian Warehouse in Upperhead Row who had the intention of selling the apparatus designed by the inventive Mr. Mori. Indeed, Mr. Robinson went as far as installing Mr. Mori's equipment in his shop for Christman 1882, the system consisting of a dynamo used to periodically charge batteries supplying six 16 c.p. incandescent lamps via fused circuits, to give a bright white light (*Yorkshire Post*, 26 December 1882). The lamps had apparently been designed by Mr. Mori who also invented an arc lamp which had its first public showing on 14 March 1883 in large sheds belonging to Oatland Mills, Meanwood Road (*Yorkshire Post*, 15 March 1882).
18. *Leeds Mercury*, 29 September 1882.
19. *Ibid*, 10 November 1882.
20. *Yorkshire Post*, 14 October 1882.
21. *Ibid*, 4 December 1882.
22. *Leeds Mercury*, 6 December 1882.
23. Leeds Corporation Report on Electric Lighting, by Sub-Committee April 1883.
24. *Yorkshire Post*, 25 April 1883.
25. *Ibid*, 3 May 1883.
26. *Electrician*, 26 May 1883.
27. 'Supplemental Particulars relating to the proposed lighting by electricity of portions of the Municipal Buildings at Leeds for the Corporation', issued by the Town Clerk, 29 May 1883, published in *Electrician*, 2 June 1883.
28. *Electrician*, 27 January 1883.
29. *Ibid*, 12 May 1883.
30. *Yorkshire Post*, 15 September 1883.
31. *Electrician*, 13 October 1883.
32. *Ibid*, 27 October 1883.
33. *Ibid*, 10 November 1883.
34. *Ibid*, 7 December 1883.
35. *Ibid*, 17 April 1884.
36. *Ibid*, 19 June 1885.
37. *Ibid*, 31 January 1885.
38. *Ibid*, 16 February 1884.
39. *Ibid*, 31 January 1885.
40. *Ibid*, 10 February 1886.
41. *Yorkshire Post*, 17 September 1883.
42. Jackdaw, 15 September 1883.
43. *Yorkshire Post*, 10 October 1883.
44. *Ibid*, 14 September 1883.
45. *Ibid*, 18 September 1883.
46. *Ibid*, 10 October 1883.
47. *Punch*, quoted in *Leeds Mercury*, 9 October 1883.
48. *Yorkshire Post*, 6 October 1883.
49. *Ibid*, 2 October 1883.
50. *Ibid*, 10 October 1883.
51. *Ibid*, 10 October 1883.
52. Jackdaw, 13 October 1883.
53. *Electrician*, 20 October 1883.

54. *Yorkshire Post*, 11 October 1883.
55. *Electrician*, 20 October 1883.
56. Jackdaw, 13 October 1883.
57. *Yorkshire Post*, 10 October 1883.
58. *Leeds Mercury*, 9 October 1883.
59. *Yorkshire Post*, 15 October 1883.
60. The enterprise was set up by the National Telephone Company, whose exchange was in Commercial Buildings, Park Row, and who organised what must surely have been one of the first outside broadcasts in England (*Yorkshire Post*, 12 October 1883). They established communication between the Town Hall and their HQ by means of five circuits and ten transmitters, inside the Victoria Hall. There were numerous applications from the 200 subscribers to the Company to be allowed to listent to the concerts, but these had to be limited to 25 at a time. Despite this limitation, subscribers in Bradford, Dewsbury, Huddersfield and other surrounding towns, as well as a small audience in the Manager's Office, were able to enjoy the privilege – a lady in the library of her house in Headingly Hill was said to have heard all of a concert with perfect distinctiveness (Jackdaw, 13 October 1883).
61. *Leeds Mercury*, 19 October, 1883.
62. Jackdaw, 20 October, 1883.
63. *Leeds Mercury*, 27 October 1883.

Chapter 7

The Municipal Buildings

Wasteful blundering

The Council's Annual Statutory Meeting in 1883 was held on Friday 9 November, less than a month after the successful display of the electric light in the Victoria Hall, which incidentally had been the only apparent sign of activity by the ELC for a long time. The fact that even this activity had been achieved virtually by the sole efforts of the Mayor and Mr. Greig caused the Council to question the capability of the ELC, and during the annual appointment of this committee a discussion arose in which the retiring committee were accused of wasteful blundering. Two or three Aldermen wished to retire from the ELC but the Council urged that they should not shirk their responsibilities, and refused to accept the retirements.

The ELC Chairman, Ald. Tatham, tried to justify the actions (and inactions) of his committee by pointing out that only £494 had been spent, irrespective of £200 for lighting the Victoria Hall.[1] In a determined effort to prove its capability and to provide electric light in the Municipal Buildings, the ELC appointed a sub-committee on 21 November to investigate the means of generating power, a sub-committee which reported on 18 January 1884.[2]

The sub-committee began by saying that their main interest had been to discover the relative merits of gas and steam engines. Mr. Crompton had written deprecating the employment of gas engines, asserting that this would result in disaster. It was true that gas engines had certain disadvantages in large installation, in that they could not perform beyond their normal power, as steam engines could, but Crossleys claimed that their 'Otto' engines were used successfully in many places for electric lighting. In view of the conflicting claims, categorical questions were sent to both Cromptons and Crossleys, who both returned very full answers.

The sub-committee also sent letters to many of the Otto gas engine users and received courteous replies from Lord Salisbury (Hatfield); Lord Randolph Churchill and Mr. W.S. Gilbert (London); South Kensington Museum, the Society of Arts and the Royal Society (London); Prince of Wales Theatre (Birmingham) and the Theatre Royal; the Mansion House of London; and the Grand Hotel and Willis's Rooms (London). The replies indicated that gas engines were advantageously employed for small installations and where there was a lack of space, but steam engines were better for large installations, as gas engines were not able to work as steadily. There was also a noticeable absence of breakdowns with gas engines, the few that did occur normally being traceable to the man in charge of the engine.

The sub-committee also reported that it had visited an Industrial Exhibition at Huddersfield but had been unable to obtain any extra information. As a result of their investigations, the sub-committee made the following recommendations:

1. As space for steam engines, boilers and associated fuel is hard to find near the Municipal Buildings, gas engines are to be used. This necessitates a reduction in the lighting load, so the electric light is to be applied only to the Library (not the Pay Office).
2. Previous plans to be used (except for the Pay Office) but with more lamps.
3. Only a limited number of firms be asked to tender, i.e. Fyfe–Main, Jablochkoff, Pilsen Joel, Ferranti, Swan, Paterson and Cooper, Anglo-American Brush, Crompton–Winfield, and the Electric Sun Lamp Co.
4. To use two (or three) Otto gas-engines (Messrs. Crossley) in the basement of the Municipal Buildings.
5. The Corporation to purchase the plant and machinery outright on successful completion of work.
6. The sub-committee be instructed to proceed to tender, and open and tabulate tenders for the main committee to decide.

The ELC accepted these recommendations by eight votes to seven and immediately set to work.

A month later, the *Yorkshire Post*[3] noted that two Crossley Bros. 12 h.p. Otto gas engines had already been purchased by the ELC and would shortly be installed in the basement. It was also reported that the ELC had accepted the tender by Messrs. Paterson and Cooper, of St. Pauls Telegraph Works, London, for lighting the Library portion of the Municipal Buildings at a cost of £1,500. 284 Swan lamps were to be used, 140 in the Reference Library, 74 in the General Reading Room and 70 in the Lending Library,[4] and the job was to be finished within two months of acceptance of the tender. It was hoped, therefore, that the electric light would be available by 1 May.

When the significance of this decision was realised, and that an earlier date was impossible, the unfortunate ELC found themselves at the centre of another row. It had only been four weeks previously that the Purchase of Property Committee had announced that Mr. Edwin Woodhouse, serving his second year as Mayor, was to open the new Buildings on the Thursday of Easter Week,[5] accompanied by a distinguished company which would include the Secretary of State for the Home Department.[6] In other words, the electric light would not be ready in the Municipal Buildings until a month after their opening.

The *Yorkshire Post* was indignant, an editorial on 25 March commenting:

'Remembering what happened at the Musical Festival, it is a wonder that the

ELC have not shown more public spirit. At that time, the slowness of the Committee excited so much indignation the matter was practically taken out of their hands by the Mayor and Mr. Greig. It would seem almost that the Committee are courting another humiliation. If the new buildings are to be lighted by electricity on the 17th prox. when a distinguished company will be guests of the Mayor, the Committee must either bestir themselves, or stand aside and again allow some more public-spirited gentleman to step in, and show them how Leeds likes its affairs managed.'

As always, Jackdaw was less restrained, writing on 29 March:

'The blundering of which the ELC seem to have been guilty in connection with this matter is nothing short of a scandal. The Committee positively seem to have gone out of their way to ensure that the work should not be done in time. It would be interesting to know what excuse they can offer for that which, upon the surface, appears to be so gross a neglect of duty.'

At the April meeting of the Council[7] the Mayor was accused of leaking information to the newspapers in a deliberate attempt to cause trouble for the ELC. This he denied, but retorted that the paragraphs in the papers were true. He pointed out that the original date fixed for the opening of the Municipal Buildings had been July 1883, but this had been delayed – at the request of the ELC – to September. Again, the ELC had been unable to supply light, and the opening had been further postponed until this month, the ELC having given another assurance that they would be ready. Now he was told that again the ELC could not complete in time. He did not know what fault could be found in the newspaper reports.

There were counter-accusations that the three members of the Gas Committee who were also members of the ELC had acted as stumbling blocks. Ald. Bower (one of the three) in denial said that he was convinced that they would all be glad to see lighting by electricity pushed forward with success. If the Municipal Buildings were a success he would support permanent introduction into the Victoria Hall.

The only result of the various arguments was the realisation that the ELC, having let the contract for the electric light, were not willing to put themselves out any more. As far as they were concerned, there would be no electric light at the opening of the Municipal Buildings, either there or in the Victoria Hall.

The opening of the Municipal Buildings

The Mayor had other ideas, however. He had arranged a banquet for 400 men on the evening of the opening and was determined that the electric light would at least be available in the Victoria Hall for his banquet.[8] The opening had been arranged for Thursday 17 April, 1884, and a bare twenty-four hours before, the Mayor asked Crompton-Winfield once more to help him out. Fortunately Crompton's fittings and engine has not yet been removed from the Victoria Hall after the Music

Festival – there had been much debate as to whether the Corporation should buy the installation – so Cromptons were able to re-commission the lighting quickly and easily, work which Mr. Crompton freely gave as a personal contribution to the ceremony.[9] The result was lighting which worked admirably, according to the *Yorkshire Post*, 'adding in no small measure to the brilliance of the scene and the pleasure of the Company.'[10]

The success of the opening must have been a great relief to the Mayor, and no doubt most of the inhabitants on what was after all a day of great pride. Leeds was a town of growing importance and prosperity, and the Municipal Buildings had

Fig. 15. *Leeds Town Hall (engraving in Leeds City Libraries)*

been built with a flourish. 'In no other town in this country', challenged the *Yorkshire Post*,[11] 'are Municipal offices on a scale of equal magnitude so completely concentrated as they now are in Leeds, or so harmoniously designed.' Indeed the external architectural features of the offices were considered too beautiful to be hidden from view, and no flags or bunting were hung outside.[12] The opening was the culmination of more than five years' work, the foundation stone having been laid on 14 October 1878. But for the delay by the ELC the buildings would probably have been opened a year earlier.

Two months after the opening the electric light was ready in the new building and the old library in Infirmary Street was able to close. 30,000 volumes were moved over to the new Lending Library and 40,000 volumes to the Reference Library, in two hectic days.[13] The new light was reported to be clear and steady.[14]

'I was glad to hear that the electric light is a great success' wrote Jackdaw, 'and that the officials of the Library are already conscious of an improvement in their own physical condition, in consequence of the exchange they have made from the foul atmosphere of the old building into the cool and well ventilated rooms in

Fig. 16. *The New Municipal Buildings (engraving in Leeds City Libraries)*

which they now have to discharge their duties.'[15]

The lighting was by 284 incandescent lamps, supplied from two improved Paterson and Cooper dynamos, self-regulating so that groups of lamps could be switched on or off without affecting the others. Under normal conditions, both dynamos supplied half the load, the circuits so arranged that in the event of the stoppage of either dynamo no part of any room would be without light. Provision was also made to work both circuits from one machine if it became necessary. The dynamos were positioned in the basement, as were the engines which drove them. There were two engines, both self-starting Otto gas engines of 12 h.p. nominal power, supplied by Crossleys of Manchester. To provide steadiness of working – difficult with gas engines – both engines were fitted with two heavy flywheels, the

Fig. 17. *The Mayor opening the new buildings 1884 (engraving in Leeds City Libraries)*

dynamos with one. The driving gear was not attached rigidly, but by a spring arrangement to modify the impulse of each combustion of gas. The tightness of the driving belts was controlled by double ratchet screws holding the machines to the frames. The mains and wires used in the work were described as being of the best possible character, and the electricity was controlled by a very handsome switch-board, provided with safety fuses, main switches, coupling switches and instruments for measuring the current and its electro-motive force. In addition, fifteen others were placed in lock-up boxes in the various rooms. The pendants, brackets and other lamp fittings were specially designed, and made by Messrs. Charles Smith and Sons of Birmingham.[16]

Disappointment and litigation

There was great disappointment when, five months after its installation, it became obvious that the electric light installation was not the success that had first been claimed. The *Yorkshire Post* reported, on 11 December 1884, that gas lamps were having to be installed, partly to give light to cleaners in the morning before the electric lamps were switched on and partly because the electric light had lately been

Fig. 18. *Municipal Buildings – plaques commemorating foundation and opening*

of a poor character, owing to a difficulty in renewing the lamps. It soon became apparent that not only the lamps but the dynamos were of poor quality, and not able to adequately supply the necessary load.

The Corporation had fortunately not completed payment, and refused to do so. The contractors initiated litigation, so that a hearing was organised at the York Assizes for early summer, 1885. The Corporation retained Mr. Webster, QC, Mr. Lockwood, QC, and Mr. Moulton. The Chairman of the ELC thought that his Committee had been badly treated by the contractors, but there were many who did not believe in the competency of the ELC and thought it must surely be the ELC's inadequacy which had caused the problem. When the ELC minutes were moved for adoption at a Council Meeting on 6 May 1885, one Alderman asked what the ELC was doing and what it hoped to accomplish – an anonymous voice said 'least said, soonest mended.'[17] Jackdaw, of course, held little regard for either the Council or the ELC, who he claimed had 'wasted the funds of the ratepayers in abortive attempts to introduce the electric light into the Library, and in the litigation consequent upon the failure of their foolish plans.'[18]

However, on 29 July, the Town Clerk reported a satisfactory end to the dispute at a special meeting of the ELC.[19] He first reminded members that Paterson and Cooper had claimed £760.10s.0d (£500 alleged still due and £260.10s.0d extras), then detailed the settlement he had made with the plaintiff's solicitors:

1. Proceedings stayed. All charges of fraud by the Corporation unreservedly

Fig. 19. *Interior of Town Hall, showing both incandescent lamp chandeliers and arc lamp globes (engraving in Leeds City Libraries)*

withdrawn.
2. The present dynamos to be removed by Paterson and Cooper and replaced with two new machines.
3. Dynamos to be tested at the Works by Sir Fred. Bramwell (President of the Institution of Civil Engineers) on behalf of the Corporation.
4. Paterson and Cooper to provide and fix 284 Swan or other approved 20 c.p. (90–100 V) lamps at their own expense.
5. The claim for extras to be referred to a 'gentleman of standing' to be agreed upon in Leeds or the neighbourhood.

6. £500 to be paid by the Corporation once the new dynamos are working satisfactorily. This is to be before 1 October.
7. Paterson and Cooper to run and maintain the installation for 14 days at their own cost.

The settlement was agreed to be very satisfactory, having saved the Corporation considerable legal expenses, particularly now that Sir Richard Webster had recently been appointed Attorney-General, and Mr. Moulton had become a QC.

In December the *Electrician* reported that the electric light had been running with steadiness and brilliancy in the Leeds Library for the last two months. 'The Free Library officials speak in the highest terms of it, and it is in this department where the most severe test is made. It is several c.p. better than the last installation, and at the same time it does not dazzle the eyes. So far as can at present be judged, the Corporation of Leeds have as good a system of electric lighting as there is in the Country.'[20]

Notes and references

1. *Yorkshire Post*, 10 November 1883.
2. *Ibid*, 19 January 1884.
3. *Ibid*, 28 February 1884.
4. *Electrician*, 15 March 1884.
5. *Yorkshire Post*, 26 January 1884.
6. *Ibid*, 14 March 1884.
7. *Ibid*, 3 April 1884.
8. *Ibid*, 16 April 1884.
9. *Electrician*, 26 April 1884.
10. *Yorkshire Post*, 18 April 1884.
11. *Ibid*, 17 April 1884.
12. *Ibid*, 16 April 1884.
13. Jackdaw, 21 June 1884.
14. *Yorkshire Post*, 6 June 1884.
15. Jackdaw, 19 July 1884.
16. *Yorkshire Post*, 16 April 1884.
17. *Ibid*, 7 May 1884.
18. Jackdaw, 4 May 1885.
19. *Yorkshire Post*, 30 July 1885.
20. *Electrician*, 18 December 1885.

Chapter 8

Other installations

Despite the passing of the Electric Lighting Act in 1882 the spread of the electric light proved to be disappointingly slow. Even the Town Council, with all its resources, had failed to illuminate the streets or give a public supply and were now having problems in providing the electric light to the whole of their own Municipal Buildings. It must have been with some embarrassment, then, that they learned of a large installation in the town which not only seemed to be successful and adequate but which was first demonstrated in the presence of Royalty.

The Coliseum

The Yorkshire College of Science had been founded in 1874 'to supply instruction in those sciences which are applicable to the Manufactures, Engineering, Mining and Agriculture of the County of York; also in such Arts and Languages as are cognate to the foregoing purpose.' Eleven years later grants to the College enabled it to build new premises off Woodhouse Lane, on the site of the present Leeds University, which the College became by charter in 1904.[1] The College Council considered themselves fortunate that the Prince and Princess of Wales had agreed to perform the opening of these new buildings, an occasion which would surely add to the status and reputation of the young College. It was essential, of course, that the Royal visit was a success, and one problem encountered was that of finding a suitable place in which the Royal couple could be entertained to luncheon. So many people wished to be present at the Prince's lunch that there was the possibility of not finding a suitable hall in which they could all dine together. Even the Town Hall was considered too small.

It was fortunate that at that time the Coliseum in Cookridge Street, which contained a beautiful, vast hall, was nearing completion. This building had a strange history, as it had originally been a wooden circus, or coliseum, which by 1882 had deteriorated to such an extent that it had become an eyesore and a public nuisance. The site was ideal for public use, being near the centre of Leeds and only a stones throw from the Town Hall, so the wooden structure was demolished and a new building started.

By chance a few months later, the touring American evangelists, Moody and Sankey, wanted a large area for their meetings, so a roof was hurriedly put on the unfinished building, and galleries built. Successful meetings were held for two or three weeks. After that there was very little progress on the still incomplete building, as the directors did not really have a clear idea about the eventual use of the building. It was only when the company was taken over by a new set of directors that progress was renewed, and it was another religious meeting that revived the fortunes of the Coliseum. At a revivalist rally it was discovered that the acoustics of the hall were surprisingly good, indeed so good that the hall was said to be the finest large hall in England. And it certainly was a large hall, able to hold 4,500 people, compared to the 1,900 who could be seated in the Town Hall.[2] It

Fig. 20. *Yorkshire College, about 1890 (engraving in Leeds City Libraries)*

was clear then that the Coliseum would be ideal for large concerts, and indeed it was successfully so used for many years. Bands and orchestras from all over England performed there for many years. There was an appearance in February 1888, for example, of Mr. Hallé and his 'augmented band' of over a hundred performers.[3]

When the College directors discovered that the Coliseum was almost finished, it was obvious to them that it would be the ideal place to accommodate the Prince and Princess, plus the 500 guests and 4,000 visitors, for the lunch on 15 July

1885.[4] The fact that the hall was to be lighted by electricity was even more impressive.

A private road ran next to the Coliseum, between Cookridge Street and Portland Crescent, and under this road was a basement room which contained the boiler, engine and dynamos.[5] The plant consisted of a 120 lbs per square inch boiler, a semi-portable 25 h.p. engine (maximum 80 h.p.) and two Mather and Platt dynamos, of 93 per cent efficiency. The building was completely lighted throughout

Fig. 21. *Coliseum, Cookridge Street, Leeds; later the Gaumont Cinema, now Leeds Playhouse Wardrobe Store. Note proximity of Town Hall on the left*

by seven 800 c.p. arc lamps and 265 incandescent lamps in ground glass globes. The lights were connected alternately to either machine, so that loss of one dynamo caused loss of only half (and alternate) lamps. Each half – and each lamp – could be switched separately. The contractors, Messrs. Cardner, Allen and Co. of London, believed there was enough power available to supply adjacent public buildings, a claim which must have been embarrassing to the ELC who were consistently having difficulties in lighting all of their own building.[6]

At this time the Yorkshire College did not have the electric light, but circumstances soon dictated otherwise. Only five months after the Royal opening of the new premises the heat from the gas 'sunlight' lamps caused a fire in the roof of one of the new rooms, persuading the College Council hurriedly to install the electric light in the Chemistry lecture room, the Physics lecture room, the Engineering lecture room and the College Library.[7] This installation was finished on 12 January 1886. Also, when the new Engineering Department was completed, ready for the October term in 1886, the whole of this building had the electric light.[8]

The Victoria Hall

The fire at the College caused the Leeds Council considerable anxiety, as the same gas lighting was in use in the Victoria Hall and some of the Municipal offices, as well as the Grand Theatre, the Albert Hall and numerous other public buildings.[9] However, Mr. George Bray, a Leeds gas lighting Engineer, assured the Corporate Property Sub-Committee that the corporation sunlights were perfectly safe. There was no need to do away with the gas lights.[10]

Despite this assurance the electric light still remained in the Victoria Hall where it had been left since the Music Festival of 1883, despite attempts by the ELC to remove what had, after all, been intended as a temporary installation. Each time removal had been attempted, immediate adverse publicity had persuaded them to postpone. When they once more proposed removal, in early 1885, it was this time the Committee of the Leeds Music Festival who objected, and strongly enough for the Council to invite a deputation from the Music Festival Committee to meet the full Council at its meeting of 6 May. The deputation told the Council that they deprecated the proposal to put back the old gas sunlights, which caused great draughts in the hall as well as producing an amount of heat which produced serious inconvenience and was almost unbearable. They argued that anyone with experience of both forms were much impressed with the superiority of the electric light, which had been very greatly admired at the last Festival.

The Council replied that it was not right that unsightly fixtures should remain just to provide electric light for the Music Festival for one week every three years, but the deputation thought that the sunlights were surely much more unsightly and inconvenient.

The Mayor, now Mr. Bower, appeared to sympathise with the Music Festival deputation and told them that the Council had always tried to meet the wishes of the Festival Committee and would seriously consider the matter at the proper time.[11] It was obviously his influence which prevented the removal of the installation from the Victoria Hall, which annoyed some ELC members who complained at the June Council meeting that by some 'backstairs' influence the ELC had been persuaded to leave the apparatus in despite a decision to remove it. Ald. Woodhouse hoped the apparatus would at least remain until after the visit of the Prince of Wales in July.[12]

Crompton's installation had still not been recovered by the Annual Council meeting of November 1885, when it was suggested that there was the probability of a permanent new installation in the Town Hall.[13] On 23 December the ELC announced that it was to light the Victoria Hall with arc lamps at a cost of £450. It was pointed out that the gas engines in the Municipal Buildings were idle during the day (when the electric light was not in use) so the arc lamps would run from batteries charged up during the day. Thus the power output could be increased without addition to the generating equipment. A small sub-committee had visited Manchester and had decided on the use of eight Brockie–Pell arc lamps,[14] each of 2,000 c.p., enclosed in 22 inch diameter opal globes. The lamps had 2 foot carbons which would burn for 18 hourse, and were surmounted by automatic regulating gear said to be simple in design and excellent in workmanship. Each lamp was to be fixed on a pulley so that it could be lowered to the floor for maintenance. The 45 batteries – in the Crypt – were 500 amp. hour capacity 'L' type made by the Electric Power and Storage Co. and both lamps and batteries were to be supplied by Rowbotham and Worsley of Manchester. Supply to the Town Hall would be by cables laid through an existing subway from the Muncipal Buildings.[15]

The Leeds Borough Engineer, Mr. Hewson, MICE, had designed the system and directed its installation by the Corporation, starting in January 1886. The Crompton electroliers were still up, of course, so only two arc lamps were fitted at first, so that the ELC and the Council could judge their effect.

On 8 March the installation was sufficiently advanced for the ELC to make tests, which were completely satisfactory – the light was perfectly steady and there was an absence of noise. Encouraged, the ELC approved the installation of the other six lamps – without disturbing the Crompton electroliers – so that the Music Festival Committee could compare the two systems. They pointed out, however, that if the incandescent lamp system was used, the Music Festival Committee would have to pay extra for the special power supply which would be required.[16]

In July the ELC and the Music Festival Committee met in the Town Hall to test the lighting. The Music Festival Committee agreed that for most occasions the arc lamps were admirable, and certainly a great deal cheaper to run than the incandescent lamps, but would produce glare in the eyes of the people in the balcony. They were unanimously in favour of the incandescent lamps[17] and the next couple of months were spent arguing over terms with the ELC, who eventually agreed to make available the Corporation's equipment (although they were not sure if the dynamos would be powerful enough).[18] There was to be no daylight running of the Corporation's equipment, though, so the Music Festival Committee asked Mr. George Bray to provide a small amount of gas lighting. This he did by erecting a new form of gas light under the gallery, giving excellent light for a small amount of gas and little heat. He also provided gas lighting at the back of the orchestra, where a bright electric light would have inconvenienced the choristers.[19]

It appeared, then, that for all the ELC's efforts, the lighting for the 1886 Music Festival was poorer than it had been three years previously, and it is not surprising that once again the appointment of the ELC was queried at the Annual Council Meeting. One Councillor wished to know if it was worthwhile re-appointing the ELC. Mr. Cooke replied that the ELC had proceeded very carefully and judiciously and had got full value for money so far expended. The ELC Chairman added that they had saved the town much money – dynamos which five years before had cost £2,000 were now £200. The Committee was appointed.[20]

For the next twelve months the ELC were surprisingly free of controversy, largely due to the fact that they didn't do much. This was probably the reason that Mr. Wilson at the next Annual Council Meeting in November 1887 moved rejection of the election of the ELC, contending that 'the Committee had absolutely done nothing to carry out the object they were to carry out, namely, to light the town or some portion of the town with electricity. All they had done was to squander £3,965 upon putting up an installation at the Town Hall and the expensive building opposite – an installation which if it had been put up by a private individual would not have cost half as much.' However, the ELC was, as always, re-elected, and was thus able to undertake its third and last major lighting scheme.[21]

The Fine Art Gallery

The Council had several months previously decided to erect a Fine Art Gallery and Museum next to the Municipal Buildings, and in March 1887 a sub-committee had accepted the tender of Messrs. Craven and Umpleby, New Briggate, for the erection of the new buildings at a price of £7,979, plus £924 for alterations and additions.[22] The Free Library Committee agreed with the ELC that the new Fine Art Gallery should be lighted using incandescent lamps (as in Manchester).[23] It became obvious that a large new installation would be required to supply the electric light to all the civic buildings in one combined system, particularly as the ELC decided at the beginning of 1888 to install as soon as practicable 100 incandescent lamps in the Mayor's Rooms in the Town Hall.[24]

So the ELC investigated equipment suitable to install, particularly engines and dynamos, and it was at this point that the ELC made one of its typically inept decisions, the sort of decision which in the past had caused so much derision and criticism. In February 1888 they decided to use compressed air engines to drive the new dynamos and accepted an offer from the Leeds Compressed Air Company to supply five 25 h.p. engines at a rental of £50 p.a. for three years.[25]

On the face of it, this was an eminently sensible decision, the choice of compressed air motors providing safety against fires and the avoidance of boilers, and obviating the need for tall chimneys. The difficulty lay in the fact that the Leeds Compressed Air Company was not actually in a position to supply compressed air to the Corporation, being one of two 'power' companies set up in 1886 (the other being the Leeds Hydraulic Power Company) who had had good intentions but no real appreciation of the difficulties involved in their business.[26] Both companies were similar in that they both wished to supply power to customers' premises from central power houses via mains laid in the streets. One company was to supply compressed air to power motors at points of work, the other was to supply hydraulic power for similar use, but neither company seem to have appreciated the vast sums which would be required to lay their mains round the town, nor the fact that there would be but limited use for their services, and therefore only poor return on capital. As much as four years later, on 9 May 1890, the *Yorkshire Post* was reporting that the Leeds Compressed Air Company was not yet ready to go ahead due to their financial difficulties.

It is difficult to say why the ELC took their strange decision. The Committee had spent much time during its short life in refusing to accept any technology until it was well proven and now here it was accepting without question a system not only unproven but uninstalled and doubtful. Fortunately the ELC had about seven months in which to provide the electric light, the opening being fixed for 3 October 1888,[27] and they realised in time that the compressed air motors would not be available. They set up a joint committee with the Fine Art Committee[28] to co-ordinate the electric lighting work and arranged inspection of likely equipment, at the end of June visiting the Leeds Works of Greenwood and Batleys to look at their dynamos.[29] They decided against these machines, however, not considering them efficient enough, and chose Edison–Hopkinson dynamos, made by Mather and Platts of Manchester (although at £2–3,000 for the four they were approximately 50 per cent more expensive.[30] The dynamos were driven by four Willans high-speed engines, made in Leeds, and the whole installation was erected in the basement of the Municipal Buildings. Of the four sets, two were required to light the Fine Art Gallery and two were used for the Victoria Hall, the Mayor's Rooms, and the rest of the Municipal Buildings.

One thousand 16 c.p. Edison incandescent lamps were used in the new buildings although the Central Hall was illuminated by six powerful 500 c.p. lamps and three more were installed in the large picture gallery. The cables were Tatham's patent lead covered and the switches were to Wilson Hartnell's patent design. The whole system was installed under the direction of the Borough Engineer, Mr. Hewson.[31]

It was pleasant that the ELC managed to complete the installation on time, for a change; in fact the new Reading Room in the Art Gallery was open and lit by incandescent lamps at the beginning of September and was able to open at night.[32] Long hours of opening were contemplated for the Fine Art Gallery, too, a prospect which excited the *Yorkshire Post*: 'The adoption of the electric light throughout the building will be of great advantage as it will permit of the Gallery being open at night – when it will be visited by the working classes – without the risk of danger to the pictures that would exist if gas were employed.'[33]

The Mayor – Ald. Scarr – opened the new buildings without incident on the afternoon of 3 October, a ceremony followed by an evening Conversazione. The ELC must have felt that at last the Committee would be elected without incident. Not so, though, for at the Council's Annual Meeting in November an amendment to a proposal moved that the ELC not be appointed. It was claimed that only £4,424 had been spent of the £10,000 allocated and only the Municipal Buildings had been lit up; if the money had been properly spent the Committee would now be providing the Town with light.

However, the ELC Chairman (Ald. Tatham) this time had no fear of criticism – he held a trump card and proudly announced that the ELC were about to offer to supply the public![34]

This was undoubtedly what the Council (and the Leeds public) wanted to hear, but why the sudden change of attitude by the cautious ELC? The answer was a simple amendment to the Electric Lighting Act 1882. After six years of argument Parliament had at last passed the Electric Lighting Act of 1888.

Notes and references

1. 'An Outline History of the University of Leeds', E. J. Brown.
2. *Yorkshire Post*, 1 July 1885.
3. *Ibid*, 23 February 1888.
4. *Ibid*, 1 July 1885.
5. *Ibid*, 1 July 1885.
6. *Ibid*, 16 July 1885.
7. *Ibid*, 16 December 1885.
8. *Ibid*, 31 August 1886.
9. *Ibid*, 16 December 1885.
10. *Ibid*, 1 January 1886.
11. *Ibid*, 7 May 1885.
12. *Ibid*, 5 June 1885.
13. *Ibid*, 10 November 1885.
14. *Ibid*, 24 December 1885.
15. *Electrician*, 1 January 1886.
16. *Yorkshire Post*, 9 March 1886.
17. *Ibid*, 22 and 28 July 1886.
18. *Ibid*, 11 September 1886.
19. *Ibid*, 9 October 1886.
20. *Ibid*, 10 November 1886.
21. *Ibid*, 10 November 1886.
22. *Ibid*, 10 March 1887.
23. *Ibid*, 13 December 1887.
24. *Ibid*, 27 January 1888.
25. *Ibid*, 25 February 1888.
26. *Ibid*, 6 May 1886.
27. *Ibid*, 15 June 1888.
28. *Ibid*, 2 June 1888.
29. *Ibid*, 26 June 1888.
30. *Ibid*, 2 August 1888.
31. *Ibid*, 28 September 1888.
32. *Ibid*, 4 September 1888.
33. *Ibid*, 28 September 1888.
34. *Ibid*, 10 November 1888.

Chapter 9

Electric Lighting Act 1888

Difficulties within new industry

The electric light did not spread at the rate that had been expected after the passing of the Electric Lighting Act of 1882, in fact many of the electric lighting companies which had been set up at that time of great optimism went into liquidation, just as Yorkshire Brush had done.

There was, naturally, a general sense of disappointment and pessimism which was reflected in the *Electrician* when it wrote: 'the contrast between the financial prospects and predictions of 1880–81 and the sober certainties of 1884 causes many of those who in earlier periods were full of confidence to speak with mournful assurance of the 'collapse of electric lighting."[1]

There were frequent allegations that it was the Act itself which had killed off the possibilities of electric lighting, but there is little doubt that the industry's troubles were caused not by the Act but by a lack of technological knowledge and expertise, referred to by the Leeds Agent of Hammonds in a letter to the *Electrician*.[2] He pointed out that the Act erroneously assumed (1) that a central station was not only feasible but practicable (actually there were *none* except for Edisons at Holborn Viaduct) and (2) that the construction of underground systems was understood (actually experience was almost nil).

The *Electrician* agreed.[3] 'Where was the experience in the companies,' it asked. 'Experience there was none!'

According to a Professor George Forbes, MA, FRS: 'the greatest difficulty of an engineering character which was encountered in the early attempts to introduce central station electric lighting lay in the enormous cost of the copper conductors required to carry to current. This, as much as the Electric Lighting Act of 1882, was the cause of the abandonment of all the earlier schemes.'[4]

This then was the root of all the industry's difficulties: there was neither the experience nor the technical understanding to appreciate the complexities of widespread distribution of electrical power, with the result that many of the cables had to be made enormously large and cripplingly expensive.

The problem lay in the fact that there was little realisation that as electrical power is a product of voltage and current, the voltage must be kept as high as possible, thus reducing the current, for it is the magnitude of the current which is the vital factor in the design of cables. Current flowing in a conductor produces both voltage drop and power loss along the conductor and so the conductor has to be large enough in cross-sectional area to minimise these effects. Obviously, then, the larger the current the larger the cable. A point is soon reached, of course, where the cable becomes not only financially exhausting but also physically impracticable to make, handle, or lay in the ground.

It's interesting to note that there is a recognition today in our modern supply industry that 240V supply is impracticable beyond two or three hundred yards. Distribution over greater distances is at voltages varying between 11,000 and 400,000 volts, depending upon the load to be carried and the distance, the aim always being to minimise the current carried in the conductors.

This variation of voltage is easily achieved in AC systems using 'transformers' but it is not possible with DC systems, which are restricted almost completely to using one voltage for generation, transmission and distribution. This voltage must obviously be suitable for use by the consumer and must therefore be low – currents in DC supply systems must therefore by high (unless the area of supply is very small) and cables must be large.

Unfortunately at the time of the 1882 Act many companies supported and ran DC supply systems but soon found themselves unable to overcome the problems of large currents and large cables. It's strange that even when Ferranti, among others, proved the feasibility of high voltage AC distribution with its use of transformers, many companies refused to change from DC supply and many DC systems lasted until well into the twentieth century. As late as 1925 there were still twenty-four different types of DC systems in use, by dozens of companies,[5] all displaying that proudly inefficient parochialism so typical of the British electricity supply industry!

Despite these inadequacies of the electricity companies, however, there were still those who believed that the Act itself was to blame for the collapse of the industry. There was particular resentment (by the companies) about Clause 27, which allowed purchase of an undertaking by the local authority after twenty-one years, without an allowance for goodwill. This, it was felt, had effectively prevented the setting up of private undertakings, for such companies would be in a losing position – they would either fail and lose their money in the failing, or succeed and be bought out by the Corporation. 'Nobody would start a chandler's shop on that basis,' commented the *Electrician.*[6]

Local authorities also objected to the Act, not so much for what was in it but for what had been excluded – they felt strongly that the issue of Provisional Orders for the setting up of private company undertakings within their areas should only be by their consent, and only if they did not wish to undertake supply themselves.

The new Act increases confidence

These points were argued strenuously for several years and several abortive attempts were made to have the Act amended, but to no avail. At last, however, after six years of intermittent Parliamentary activity, pressure from the public, the companies and the local authorities proved sufficient: the Electric Lighting Act 1888 received Royal Assent on 28 June that year.

The new Act was designed as an amendment to the Act of 1882 and was therefore comparatively short. There were, in fact, only four effective clauses in the new Act, the main points being:

1. Provisional Orders authorising the supply of electricity by private companies would be granted by the Board of Trade only with the consent of the local authority, unless the Board of Trade were of the opinion that consent ought to be dispensed with.
2. The period after which the local authority had the option of compulsory purchase was extended to forty-two years (unless a shorter period was specified in the Provisional Order) but there was still no allowance made for goodwill.
3. The Board of Trade were given the authority to vary, in the Provisional Order, the terms of sale to the local authorities, notwithstanding the requirements of the previous clause.
4. Several restrictions were made on the placing of cables to avoid interference with telegraphic equipment.

Although the amendments were not extensive, the old Act had been blamed so vigorously by the companies for their failure that the revision had great psychological importance and in itself increased confidence and hope of rapid success. Happily, too, the revision had come at a time when many of the technical problems were being overcome so that the passing of the 1888 Act coincided with a sudden upsurge in electrical supply business. Two reports from the *Electrician* serve to demonstrate how successful was this rise of the industry: On 2 March 1888 there was reported a reply to a question in the House of Commons which stated that since the 1882 Act, fifty-nine Provisional Orders and five Licences had been granted to companies, and fifteen Provisional Orders and two Licences to local authorities. None had been used.

On 3 January 1890 (16 months after the 1888 Act) it was reported that twenty-one central stations were operating in London and thirty in the provinces.

This sudden success of central supply systems was received with delight by an excited public, but it must be admitted that the Leeds Council were somewhat less enthusiastic. It's true that they now had more control over their area of potential supply, by virtue of their new power of veto over applications for Provisional Orders, but they realised that the increasing success and confidence of private companies would result in an increasing number of applications to give supply in the town. The public's expectations of a central supply would surely become impossible to resist.

The Council's choice

As a result, in the summer of 1888, when the Act became law, Leeds Council felt that they must take one of three alternative courses of action:

1. to consent to the granting of a Provisional Order to a private company, unlikely in view of their avowed dislike of the private monopoly of public services;
2. to undertake the supply of electricity themselves, despite their continued doubt about the efficiency and economy of electricity supply;
3. to do nothing, but nevertheless persuade the Board of Trade not to dispense with the Council's power of veto.

The decision, as with so many Council decisions, was to prove difficult to make and was to be a long time in the making.

This was not initially apparent, though, for in October 1888 the ELC seemed to favour the second option and instructed the Borough Engineer to prepare a scheme for the lighting by electricity of the centre of Leeds.[7] This caused the first delay, for the report took several months to prepare, and it was not until May the following year that Mr. Hewson's report was present to the ELC.[8] They heard that estimates and particulars had been obtained from two leading electrical firms and that it was proposed to erect a generating station on the 'Midden' (waste ground near the Markets) to supply an area bounded by Boar Lane, Leeds Bridge, Vicar Lane, Upperhead Row and Albion Street. However, the ELC were alarmed to hear that the cost was estimated at £40,000. The *Yorkshire Post* were convinced that the Council would not wish to spend such a large amount of money on what was, after all, 'just an experiment on a large scale'.[9]

The ELC adjourned their meeting for two weeks so that new proposals for the use of arc lamps could be submitted and also to find out how far the electric light could actually be delivered from the Midden.[10] However, when reconvened they were further dismayed to hear that Mr. Hewson now recommended the provision of £61,000 to set up and supply the centre of the town, business and commercial premises during the day and domestic premises in the evening.[11]

Still unable, or perhaps too frightened, to reach a decision, the ELC again adjourned their meeting, asking for the report to be printed and distributed so that they could study it at leisure and meet again in two weeks to decide finally.[12]

Accordingly on 12 June 1889 the ELC again met and having decided to include Kirkgate Market in the lighting scheme found that estimates had risen to £65,000.[13] Not surprisingly, a decision was again postponed for a week and the *Yorkshire Post* believed this delay was a preliminary to shelving the project. 'The more the committee think and talk about the proposals the less they like them.'[14] But it was wrong! On 18 June the ELC at last recommended acceptance by the Council of a scheme at an estimated cost of £68,000, to include not only the central but also the main residential area, a scheme which would required an initial outlay of £25,000 when the Provisional Order was obtained.[15]

They also recommended an extension of the electric light in the Municipal Buildings (in the Council Chambers, Courts of Justice and various offices) by means of 2,000 incandescent lamps at an estimated cost of £3,000.

A twenty-seven page report was prepared by the ELC giving not just their proposals for the future but a history of their work in the past.[16] This report pointed out that the difficulty of distance had now been overcome, so that small-section high potential mains could be used instead of large-section low potential mains. A poll of shop tenants in the city centre showed an almost unanimous desire for the electric light and in the proposed area 37,000 lamps (of 10 c.p.) would be required. £58,000 was required for the commercial part of the system and £10,000 to supply the evening (domestic) load. The cost of 1/8d per lamp per hour would be equal to gas at 3/5d per 1000 cu ft, or almost double the cost of Leeds gas.

When the Council considered the report on 3 July it agreed that the extension of lighting in the Municipal Building and the Town Hall should go ahead, but the rest was referred back to the ELC with a request for them to determine the annual cost of the existing installation and the cost in other towns of electric lighting schemes.[17] This reference back caused another three months' delay, the Council meeting in October to discuss the details which had been provided by the ELC, details which were not as conclusive as the Council had hoped. In fact the information from other towns was too varied to be of any use at all, but the ELC was able to confirm that the cost of the Leeds system (admittedly not yet finished) was £2,606 p.a. for 2,000 lamps, or .208d per lamp per hour (less than a farthing). These figures were not particularly encouraging and a heated two hour debate ensued.[18]

Decision deferred

There was widespread dismay when the Council, at the end of the debate, resolved by twenty-eight votes to seventeen to defer a decision on public supply for six months. The disappointment was deep, for it was now sixteen months since the Act had been passed and there had been continual expectations of resolutions in favour of the electric light for most of this time. It was now apparent though that the Council was actually no nearer acceptance of the electric light than it had been in June 1888.

Strangely enough a six month delay must have been a very difficult decision for the Council to make as there were several intense pressures on it to be positive, one way or the other. On the one hand there was the immense problem of lack of finance. According to the *Yorkshire Post* of 18 October the borough was one of the most heavily rated in the country, with a debt which in April 1890 had grown to the staggering sum of £4,493,839.[19] The Council obviously had no money to spare on the undeveloped and unproved science of electricity and it would have been understandable had they abandoned the idea of electric supply to private enterprise and risk.

On the other hand there were frequent references to the incompetence and inactivity of the Council, who must have been tempted constantly to prove their modernity and their acumen by investing in this new science. The public, the Press, electricity companies and even other local authorities seemed to lose no opportunity in pointing out, directly or indirectly, the ineptitude of the Council. A 'civil engineer' had already written in the *Yorkshire Post* of 7 June that 'one need only refer to the earlier experiences of the ELC for an illustration of how not to do it!'

A more sarcastic contributor to the *Yorkshire Post*, on 24 August, had declared that 'it will not do for Leeds to forfeit its conspicuous position in the rearguard of the march of progress by any such action as would expose its governing powers to the faintest suspicion of vigour.'

The *Yorkshire Post* itself, on learning of the six-month delay, issued a strongly worded editorial on 18 October 1889:

> 'No question that had been tackled by the Municipality in recent years more forcibly illustrates the shilly-shally peddling of what has come to be known as the "old school" in the Corporation than that of electric lighting. The Local Authority has betrayed a degree of feebleness that would be very amusing were it less costly and provoking.
>
> After the lighting of the Municipal Buildings and Town Hall all the allocated money had gone. "Ten thousand pounds gone, and nothing to show for it but the lighting of three or four rooms?" protested ratepayers all over the town.
>
> It is suggested in some quarters that the Corporation have all along been wittingly engaged in a combined manoeuvre that all their travel and trials are but a blind to keep electric lighting companies out of the field. But this electric lighting fiasco is no feint. Cunning has never been the weakness or the strength of our local governors. They have simply been unable to make up their minds.
>
> When (the Council) has done experimenting, travelling, reporting, debating and opposing the lighting companies, this precious Borough County Council will have spent about as much as the electric light would have cost. There is not in the United Kingdom a single town of the size, wealth and importance of Leeds that is so wretchedly lighted.'

Other local authorities were showing much more interest than Leeds, indeed the *Electrician* reported on 1 November 1889 what it referred to as a 'curious fact' – that Nottingham Corporation at their own expense had obtained a copy of the Leeds report and had decided to apply for a Provisional Order (where Leeds would not). In fact by the end of January 1890 there were forty-four applications for Provisional Orders from local authorities including such towns as Manchester, Oldham, Belfast, Birkenhead, Dover, Aberdeen and Stafford, and more locally the towns of York, Hull, Barnsley and Huddersfield.[20]

Despite this, the *Electrician* was suspicious of local authorities, suggesting (on 20 December 1889) that: 'Local authorities, in many instances, are making common cause against the applicant companies merely for the purpose of keeping the electric light out of the towns altogether. If only time enough can be gained, it is thought that the private undertakers will disappear from the field, and then the municipal applications can be quietly dropped.'

In Leeds there were initially five applicant companies: Electric Construction and Maintenance Co. Ltd.; Laing, Wharton, Downe Construction Syndicate Ltd.; Latimer, Clark, Muirhead and Co.; National Electric Supply Co. Ltd.; and Yorkshire House-to-House Co. Ltd.[21]

All of them promised supply to the centre of the town, and the Yorkshire House-to-House also promised supply in Woodhouse Lane and Headingley.[22] But the Council's attitude gave scant encouragement to the inhabitants of these areas and the frustration must have been increased by reports of the spread of the electric light in London. Jackdaw travelled frequently to the capital city and did his best to keep the Leeds public up to date with progress there. 'In nearly every street of the Metropolis the wires for the electric light are now being laid down, so that it looks as if the whole capital were under repair,' he wrote on 19 October 1889. 'By the beginning of the year it is hoped that the work will be done.' A month later (30 November) he added: 'It is evident that by the spring of next year gas lighting in London will receive a great blow, in as much as the Electric Light will be "laid-on" in almost every important street, and will be available for supply to nearly every house.'

How the inhabitants of Leeds must have looked on with envy! More and more were becoming impatient with local progress and were installing their own systems. The Trevelyan Hotel Co., for example, installed engines and dynamos in the Hotel basement (in January 1890) to supply not only the Hotel but all the block of premises in Boar Lane from the Hotel to White Horse Street.[23]

Bradford opens first municipally owned generating station

The growth of the electric light in London and the increasing number of private installations in Leeds must have been a cause of great concern to the Leeds Council, but without doubt the greatest embarrassment to the Council must have been the opening of the country's first municipally owned central electricity generating station for public supply on 20 September 1889 – just twelve days before Leeds took its six month delay decision. The reason for the embarrassment was that the station was in Leeds' close neighbour and rival – the progressive town of Bradford.

Bradford Council had been interested in the electric light for several years, and had acquired powers to supply electricity under an Act of Parliament as early as 1883.[24] In common with most of the rest of the country at that time they had eventually decided not to proceed with the electric light until it improved, a position they felt had been reached in 1887. In February of that year the Gas Supply Committee (who were to become responsible also for electric light) wrote to people in an area within a quarter mile radius of Bradford Exchange asking if they required the electric light. Receiving 111 favourable replies they then asked a well-known London electrical engineer – Mr. Shoolbred – to report on how to provide the electric light in that area and its cost.[25]

The Leeds public were not really aware of the serious intent of the Bradford Council until a report appeared in the *Yorkshire Post*, on 21 December 1887, which said that an application for an overhead road crossing by Schmidt Douglas Co. Ltd. has been turned down as Bradford had themselves decided to proceed with electricity supply. Schmidt Douglas had for some time been supplying a large block of shops and warehouses in the city centre with electric arc lighting fed on an overhead system and Bradford offered to take the installation over with the idea of using it as a base for their own system; when the offer was refused they confirmed that they were to build their own generating station on a plot of land they had already bought in Canal Road, fronting Bolton Road. This station, which would have boilers and engines, was near several large mills which would require the electric light and the underground cable system would pass such important places as the Midland Station and the Midland Hotel on its way to the city centre. The system was known to be suitable, too, for electro-platers, several of whom were wishing to set up in Bradford. It was anticipated that all those prospective customers, and any other wanting the electric light, would supply and install their own equipment, the Corporation only being responsible for supply and metering (if an accurate and reliable meter could be found).

Under the Provisional Order the maximum charge was to be 7d per unit but the expected charge would be nearer 5d per unit, equivalent to 4/2d per 1000 cu ft of gas or a little less than double the gas rate. However, many hopeful customers had said that they would willingly pay double the gas rate because of the superiority of electricity. The Council believed that the proposed system would be worth all the costs even if it were to supply only the Town Hall, Markets and Free Library.

In Leeds, Jackdaw was of course immensely disappointed! On 17 December he wrote:

'So Bradford is once more showing that in certain respects its people are more "quick" than we are in Leeds! Here are we still shivering on the brink, doubting and hesitating as to whether it would be wise or the reverse to go in for a large public scheme of electric lighting, whilst Bradford boldly prepared to make the plunge. We can hardly afford to be left behind by Bradford in such a matter as this.'

In March 1888 the Bradford Gas Committee announced that they had adopted Mr. Shoolbred's scheme. The substantial stone station was to contain three Lancashire Boilers of 140 lbs/sq in and three sets of dynamos driven direct by separate engine of inverted vertical type, the dynamos each having an output of

85,400 watts. There was provision for supplying 3,000 incandescent lamps of 16 c.p. each, supplied via two insulated and lead covered cables, one along Market Street to the Town Hall, the other through Kirkgate and Darley Street to the Markets and Free Library. However, the boiler house could be easily extended if required by the addition of another storey, so that there would be ultimate space for sixteen boilers.[26]

Permission to borrow the capital, estimated at £20,000, was given by the Local Government Board in June, and the contracts were let. Siemens were responsible for everything except the buildings and the boilers and their contract also included superintendance and management of the works for six or eight weeks after completion.[27] The Committee hoped for completion by September with lighting in full operation before winter.[28]

The works did not progress as intended, and were not helped by a fire at Whelans, the London manufacturers of the electric dynamos. The target opening date slipped to April 1889,[29] then in June it was expected to be July/August,[30] and at the beginning of September was put back two or three weeks to engage staff and work people, although the installation was tested and ready.[31]

The *Yorkshire Post* was rather patronising about the whole system. It reported that the voltage – nominally 115 volts – would be 113 volts at the lamps which would soon burn out as they were only rated at 100 volts.[32] The paper also claimed that the wrong system – low voltage DC – had been chosen. This, it argued, restricted the economical supply area in Bradford to within a quarter mile of the station. Any extension would require another generating station for every 775,000 square yards, obviously an uneconomic situation. The high voltage AC system would have been better, alleged the *Yorkshire Post*, in which electricity was able to be transmitted long distances at 2000V and then transformed down to a safe usable pressure at the premises to be supplied. It also pointed out that a low voltage 100V system required cables 400 times bigger than a 2000V system for the same power (actually it was wrong – the cables needed to be only 20 times as large, although if the cables were the same size the low voltage system produced 400 times the electrical losses).[33]

However, Bradford Council seemed happy enough with the system they had and in February 1890 announced a further expenditure of £10,000 for extensions, after only a few months operation[34]; the whole operation went well from the start and despite a first year loss of £1,200 (not unusual in these businesses) the enterprise was soon in profit.[35] By 1896 the profit was nearly £3,500 and the Council had laid the foundation stone of new works in Valley Road. A couple of years later Bradford Council was among the pioneers of hiring out equipment in order to increase the sale of electricity. Motors cost from £1.10s to £5 p.a., arc lamps were 10/6d or £1.1s. p.a. and domestic equipment ranged from water boilers at a guinea to hot cupboards at £37. The *Electrician* was impressed, commenting on 4 February 1898 that 'a copper bronchitis kettle costs £3.15s., which is no more than an average doctor's bill for curing an attack of the complaint.' It added, with approval, 'the enterprise which has characterised the go-ahead municipality of

Bradford from the first, in regard to electrical matters, in thus maintained.'

In July 1889 this success of the Bradford system was not yet apparent, but the installation did at least show that centralised supply was practicable and the Leeds Council began to be worried that the Board of Trade might be persuaded to grant a Provisional Order to one of the private companies, despite the Council's refusal to support any of the applications.

The Council reverses its stand

The accumulation of pressures on the Leeds Council at last proved overwhelming. A worried Council assembled on 28 January 1890 especially to discuss electric lighting, although only four months of their six month adjournment had passed. Signs of panic were beginning to show, and Alderman Spark sought to calm the growing fears. He explained how the electric light was rapidly becoming more economical as prices fell and pointed out that the installation in the Fine Art Gallery had fallen to a cost of £2.18s.9d per lamp against a figure of £8.14s.8d per lamp in the previous installation. He forecast that large scale production would soon produce electric lighting at the price of gas, which would surely satisfy that large number of tradespeople who were keen on the electric light. He pleaded that the six month period be allowed to expire, when the ELC could then report on the latest developments. The Council, reassured, agreed.[36]

The Council must have been even more reassured when, at the end of the six months, they received a comforting message from the Board of Trade. As the Corporation was at present involved in an experiment with the electric light, the Town Clerk reported, the Board of Trade would allow the matter of Provisional Order applications to stand for a year. However if after this time Leeds could not proceed with a proper supply, the Board of Trade would again have to consider authorising other promoters to do so.[37]

This meant that the ELC now had an unexpected twelve months in which to decide on a suitable system which it could recommend to the Council. In fact, it required only half that time. It was at the beginning of October 1890 that the ELC presented their report at a Council meeting.

It appeared at first that the ELC were to make a firm recommendation. They gave a detailed account of progress, both here and abroad, reported on costs, and then went on to state 'with absolute unanimity and strong convictions' – according to the *Yorkshire Post* of 2 October – 'that any further delay in providing the electric light for general use would be to deprive Leeds of an illuminent possessing great advantages over every other mode of lighting, especially from a sanitary point of view.'

To ensure proper discussion on a topic which was now of huge local importance and interest it was resolved to call a special Council meeting the following Monday, 6 October 1890, business to be restricted solely to the electric light, and it was at this meeting that doubt was at last cast upon the true purpose of the ELC.[38]

Doubt was not to remain long, however. The ELC Chairman, Mr. Hardwick, opened the discussion by proposing that the Corporation apply for a Provisional Order to undertake the supply of electricity in the Borough and he reminded the Council that a decision must soon be made or the Board of Trade would consider allowing a private company. Then, to the amazement of the Council, he admitted that he was actually in favour of private companies!

It must have been difficult for Mr. Hardwick to both propose and speak against the motion at the same time and it is not surprising that the Council were considerably upset by the vague and confused way in which he explained the matter. There was no improvement when Mr. Willey rose to speak, for although purportedly seconding the motion proposing Corporation control, he declared that he also preferred private companies, who, he said, looked after their business better and employed more efficient servants than the Corporation! It was argued that if the Corporation set up a system it would never get beyond the centre of Leeds, whereas private companies would wish to expand. In any case, if money was lost, it would be the ratepayers.

Considerable amazement was expressed that both the proposer and seconder had actually spoken against the motion, but it was no surprise, in the circumstances, when the Council voted overwhelmingly (by thirty-five votes to four) to let private companies do the work. In view of the previous oft-repeated statements of animosity by the Council towards private companies this decision must have been received with great astonishment by the inhabitants of Leeds and it need hardly be said that Jackdaw, for one, objected most strongly to both the breaking of this long-held principle of municipalisation and the manner of its breaking. In his column that weekend he wrote:

> 'The report of the last meeting of the Leeds County Council was neither entertaining nor edifying to read, and some of us who are ardent believers in local self government would not be able to repress a feeling of disappointment – if indeed the word disappointment will express the depth of our sentiment on the subject – by the time we had finished it. There are so many people inclined to forget in these days that government, local as well as imperial, is a serious business, and the Leeds County Council seems to have suffered a lapse in this direction. Liberty of opinion is every man's possession but when the chairman of a Committee gets up and proposes something in which he avows that he has no faith, and another member seconds the proposition in terms only perhaps a little more farcical than the first, then we must observe that the public are not being fairly dealt with. Local bodies exist to limit as far as possible private monopoly in the interests of the community, but our local administrators have interpreted their duty otherwise, and now the electric light is to be provided by private enterprise. The Council cannot undertake the risk, forsooth! In its wisdom it is going to wait until the monopoly is worth a heavy price, when, of course, it may judiciously be purchased.'[39]

Now that the Council had 'thrown up the sponge', as the *Yorkshire Post* called it, they had to start considering the Provisional Orders which three of the private companies had re-submitted to the Board of Trade, objections having to be lodged before 22 February 1891.[40] After this date the Board of Trade would be free to allow a company to go ahead or call upon the Corporation to show why it should not, and so the Parliamentary Committee were asked by the Council to investigate the claims of the three companies: Electric Installation and Maintenance Co., National Electric Supply Co. and Yorkshire House-to-House.[41]

By 13 February the Parliamentary Committee had met representatives from the three companies and had received a technical report on the Provisional Orders from the Borough Engineer[42] but even at the end of March were unable to resolve upon any definite course of action.[43] 'Is the Parliamentary Committee, which for many a week has been considering, or pretending to consider, the proposals of the three companies, dealing fairly by these companies?' asked the *Yorkshire Post* on 26 March. 'There exists at this moment a strong suspicion in many quarters that the Corporation is continuing the dog-in-the-manger policy that has been exposed in these columns again and again. At the meeting of the Committee held on Tuesday many absentees were noted, and many of the members who were present appear to have had difficulty in working up either interest in or knowledge of the business in hand.'

This may have been the spur that was required, for on Friday 17 April 1891 at a special Council meeting a resolution was proposed which was the unanimous conclusion of the Parliamentary Committee: 'That this Council, as the Local Authority for the Borough of Leeds under the Electric Lighting Acts 1882 and 1888, do hereby consent to the grant by the Board of Trade of a Provisional Order authorising the Yorkshire House-to-House Co. (Ltd.) to construct and maintain electric lines and works, and to supply electricity within the borough on such terms and conditions as the Parliamentary Committee may approve.' The motion was adopted by the Council.[44] At last!

Notes and references

1. *Electrician*, 19 July 1884.
2. *Ibid*, 5 July 1884.
3. *Ibid*, 3 January 1885.
4. Lecture printed in *Electrician*, 30 September 1887.
5. *Electrician*, 16 March 1928.
6. *Ibid*, 22 November 1884.
7. *Yorkshire Post*, 20 October 1888.
8. *Ibid*, 11 May 1889.
9. *Ibid*, 18 May 1889.
10. *Electrician*, 17 May 1889.
11. *Yorkshire Post*, 30 May 1889.
12. *Electrician*, 31 May 1889.
13. *Yorkshire Post*, 8 June 1889.

14. *Ibid*, 13 June 1889.
15. *Ibid*, 19 June 1889.
16. *Ibid*, 22 June 1889.
17. *Ibid*, 4 July 1889.
18. *Ibid*, 8 October 1889.
19. *Ibid*, 10 April 1890.
20. *Ibid*, 29 January 1890.
21. *Ibid*, 4 July, 11 November, 15 November 1889.
22. *Ibid* 28 December 1889.
23. *Ibid*, 4 January 1890.
24. *Ibid*, 1 June 1888.
25. *Ibid*, 14 February 1887.
26. *Ibid*, 30 March, 1 June, 13 June 1888 and 20 September 1889.
27. *Ibid*, 13 June 1888.
28. *Ibid*, 30 March 1888.
29. *Ibid*, 12 December 1888.
30. *Ibid*, 20 June 1889.
31. *Ibid*, 2 September 1889.
32. *Ibid*, 20 September 1889.
33. *Ibid*, 24 March 1890.
34. *Ibid*, 1 February 1890.
35. *Electrician*, 12 June 1896.
36. *Yorkshire Post*, 29 January 1890.
37. *Ibid*, 8 May 1890.
38. *Ibid*, 7 October 1890.
39. Jackdaw, 11 October 1890.
40. *Yorkshire Post*, 22 November 1890.
41. *Ibid*, 7 August 1890.
42. *Ibid,* 14 February 1891.
43. *Ibid*, 25 March 1891.
44. *Ibid*, 18 April 1891.

Chapter 10

House to House electricity

We last met Robert Hammond in 1885 when not only his own but his associated Brush electrical companies had gone into bankruptcy. The Brighton Central Station system was still going strong, though, and at the beginning of 1888 the network had grown to a total length of fifteen miles. Unit sales increased rapidly, with 15,757 units sold in the quarter ended 31 December 1887 compared with 8,795 units sold in the same quarter the previous year.[1] There was no doubt that the success of the Brighton system showed the possibilities of centralised supply, and left Hammond in the happy position of being one of the few men in the country with experience of running such a system successfully. Thus it was that on 21 February 1888 *The Times* displayed an advertisement for a new electrical supply company which was to be set up 'for the purpose of establishing and working central stations for the generation and distribution of the electric current for lighting purposes.' The company and the system were to be based on the success of the Brighton enterprise, and on Hammonds' expertise which had made this success possible. The company was called the House-to-House Electric Light Supply Co. Ltd.

House-to-House Electric Light Supply Company

If Hammond's experiences had taught him anything, it was to avoid being tied to restrictive patents and concessions, and the prospectus of the new company carefully emphasised that no works were to be taken over and no patents had been, or would be, purchased. The intention was to buy and use the most suitable equipment and pay royalties where necessary. All capital would therefore be available for the establishment and expansion of the company.

The company's intention was to set up central stations in those areas where sufficient signed contracts from customers could be obtained to justify a system (recent advertisements had produced applications for 14,000 lamps), the centralised system clearly giving advantages to both customers and company. The existing practice of each customer having his own generator was costly and

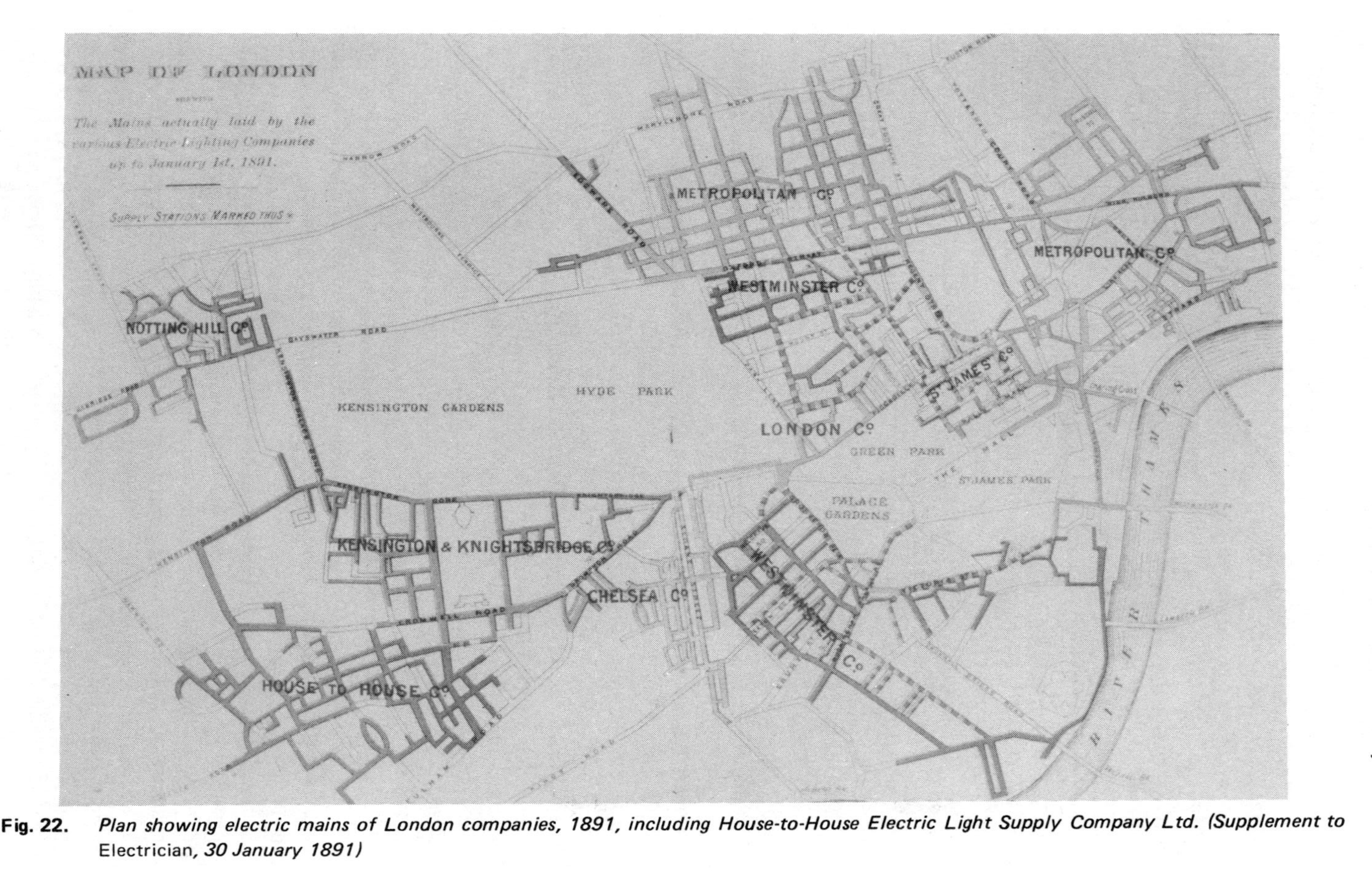

Fig. 22. *Plan showing electric mains of London companies, 1891, including House-to-House Electric Light Supply Company Ltd. (Supplement to* Electrician*, 30 January 1891)*

troublesome, with suspect reliability; a central system would produce cheaper electricity, and customers as far away as two miles could be supplied continuously and realiably. Actually, costs were expected at first to be more expensive than gas, but would fall as the systems grew and improved.

The company expected to benefit handsomely, too. If the Brighton system was taken as an example, the connection of 2,000 lights would produce a comfortable profit and there would be a proportional increase in profits with the establishment of larger plants supplying more lights.

Transmission by overhead wires was possible immediately without Parliamentary powers and, in fact, legal advice confirming this had been obtained by the new company from the Attorney General himself (Sir Richard Webster), and from Mr. Moulton, QC, and a colleague, Mr. Wright. If underground cables were needed, however, the company would be prepared to make the necessary arrangements.

The prospectus stated that the moment was now opportune for the expansion of the electricity supply system in view of the imminent new Electric Lighting Act; electric light supply companies should now make large profits just as gas companies did. The company had a nominal capital of £350,000 in 69,900 ordinary and 100 founders shares, all of £5, although only 35,000 ordinary shares were offered initially. Profits each year would provide first a 7 per cent dividend, the surplus to be divided in equal moieties between the holders of the ordinary and founders shares, thus again giving founders shares a large proportion of the profits.

Robert Hammond was to receive all the founders shares, plus £5,000 worth of ordinary shares (all fully paid up) and also £5,000 in cash. In consideration of this he agreed to be Managing Director (without remuneration) for two years and to bind himself not to associate with any other electric lighting company for five years without the consent of the House-to-House Co. To demonstrate further Hammond's involvement with the new company, the temporary offices were at his home address – 117 Bishopgate Street Within, London EC.

A year after the issue of the prospectus, the company opened its first central station, at West Brompton, London. Jackdaw felt quite justified in claiming that this was the first serious attempt to establish a system of domestic electric lighting from a central station, the first 'purpose-built' central supply system.[2] After all, both the Brighton and the Grosvenor (Art) Gallery systems had grown by accident rather than design, and the Kensington Court system was restricted to supplying a private estate.

Perhaps because of this 'uniqueness' the system grew only slowly at first. At the end of 1889 – after eleven months operation – only seventy-eight houses were connected to the system, which was supplying a total of 4,520 lamps of 35 watts (or almost 60 lamps per house on average).[3] In May 1890, though, the directors were telling the shareholders that 'every new building in the district is being wired for the electric light, and no gas pipes are being admitted at all.'[4] and indeed at the end of that year, 248 houses (with 13,665 lights) were connected to a system now extended to nine miles of mains.[5] In 1891 there was a huge 50 per cent growth so that 373 houses (19,388 lamps) were connected by the end of the year.[6]

There is little doubt that the success was due to the technical ability of the company, aided particularly by its Engineer Mr. William Lowrie (who helped design both the dynamos and the transformers) and by the Leeds firm of Fowlers, who provided the engines.[7] Before long, not only was there an increasing demand for connection to the West Brompton station but there was also a demand for the erection of other stations, abroad as well as in England (a station similar to that at West Brompton was opened in Madrid in the summer of 1890).[8] The expense of

Fig. 23. *View of part of House-to-House Central Station, West Brompton: main engines and alternators (*Electrician*, 14 November 1890)*

erecting such stations was expected to be more than the House-to-House Co. could afford, so in early 1891 a new company was set up to undertake station construction work – the Leeds and London Electrical Engineering Co. Ltd. This was a joint venture by House-to-House and Fowlers ('the celebrated firm of J. Fowler and Co. had the management of it') and Hammond was a director.[9]

It was anticipated that there would be many stations to be built, as the House-to-House Co. (and the eleven subsidiary companies which the company registered) had asked for consent to apply for 224 Provisional Orders by August 1889 – 20 in London and 204 over the rest of the country – and there were high hopes of success for many of the applications.[10]

Yorkshire House-to-House Electricity Company

One of the subsidiary companies, with 19 Provisional Order applications to its name, was the Yorkshire House-to-House Electricity Co., registered (but not capitalised) on 24 June 1889[11], and destined to spend the first two years of its life waiting in frustration while Leeds Corporation debated the pros and cons of municipal electrification. There must have been great relief and excitement in April 1891 when the Corporation at last consented to the Provisional Order application from Yorkshire House-to-House.

Not that Yorkshire House-to-House were in a position at that time to consider immediate provision of the electric light to the inhabitants of Leeds. The Corporation's consent merely signalled the start of a long period in which Parliamentary approval would have to be obtained, money found and a system built, a process begun by applying to the Board of Trade for the approval of a Provisional Order.

This first step was comparatively swift. The Leeds Electric Supply Order 1891 was approved by the Board of Trade and confirmed by Act of Parliament on 3 July.[12] The Order, a lengthy document containing eighty-two provisions and four schedules, gave details of how the undertaking was to be run, in a legal and accountancy sense rather than technically. Some provisions, for example, conferred powers enabling the undertakers to break-up streets, although with strict controls to prevent interference with equipment belonging to the Postmaster General, the gas and water undertakers, and the railway and canal companies.

Other provisions sought to protect the customer by requiring the undertakers to provide a sufficient supply which was to be properly measured with approved meters. Maximum charges for the electricity were laid down, too – 'for any amount up to twenty units, thirteen shillings and fourpence; and for each unit over twenty units, eightpence.'

The most important part of the Order, however (at least as far as prospective shareholders were concerned), was paragraph 59, headed 'Purchase of Undertaking', which detailed the agreement made between Yorkshire House-to-House Co. and Leeds Corporation. Under this agreement the Corporation had the right to take over the works and business of the Company on the following conditions:

(a) Within ten years:
  (i) by the issue to the company of such an amount of Corporation stock as would produce, by the interest thereon, an annuity of 5 per cent p.a. upon the sum properly expended by the company and chargeable to Capital Account;
  (ii) plus the payment to the company of a sum equal to the aggregate amount of a dividend of 5 per cent p.a. on the said Capital Expenditure, less the aggregate amount of the dividends declared by the company up to the date of purchase.

(b) After ten years but within twenty-one years: as in (a) (i) above, to produce the 5 per cent annuity.

(c) After twenty-one years but within thirty-one years: either under the terms of (b) above, or by paying the company the then value of the undertaking (according to Section 2 of the 1888 Act) together with a sum of goodwill to be determined by arbitration).

(d) After thirty-one years (after July 1922) or at the end of every subsequent period of seven years:
Either under the terms of (b) above, or by paying the company the then value of the undertaking (but with no goodwill). The options of purchase on the terms of (b) cease in 1933.

The area of supply was defined in the first schedule as 'the whole of the Municipal Borough of Leeds'; the second schedule listed the streets and parts of streets throughout which the undertakers must lay distributing mains within a period of two years after the commencement of the order. These lay in an area covering virtually the whole of the centre of the town, but also included Woodhouse Lane and Headingley Lane in the rich north-west suburbs of Leeds (David Greig of Fowlers lived in Headingley Lane, near the fine mansion Spring Bank, although the Kitsons had now moved to far grander premises at Gledhow Hall).

The prospectus is published

Having thus equipped themselves with powers to supply electricity in Leeds, the Yorkshire House-to-House Co. felt their next steps were to acquire land for a central station and determine the system and works necessary for a successful enterprise. This took such a surprisingly long time that many people must have believed the whole thing had fallen through. Indeed, the *Electrician* reported that managers of several leading banks and secretaries of some of the most important clubs, and others, had issued a joint appeal to Leeds Corporation asking for a supply of electricity at a reasonable rate.[13] Yorkshire House-to-House had not collapsed, however, for only a month later the prospectus of the company was published, offering shares to the public.

From the prospectus, dated 26 March 1892, it was clear that this was no 'foreign' company with a local name, for the Directors, Bankers, Brokers, Solicitors, Auditor and Secretary were predominantly from Leeds and the temporary offices were in Leeds, at 32 Park Row. Of the seven directors, five were Leeds men: Grosvenor Talbot of Burley (Chairman); Robert Eddison of Fowlers; Robert Hudson, who had a foundry at Gildersome; Samuel Ingham, a timber merchant; and Arthur Lupton, cloth manufacturer. George Crowther of Huddersfield and John Pearson of Thirsk completed the Board.

The prospectus declared that the company was to be capitalised at £100,000, divided into 20,000 shares of £5 each, of which 19,900 were ordinary and 100 were founders shares, although only 10,000 ordinary shares were on offer initially. The founders shares would receive half the profits (after the payment of a 7 per cent dividend) and had originally been allotted to the parent House-to-House Co., but that company had agreed to place them at the disposal of the directors of the new company who therefore decided to allot 50 founders shares to themselves, one share being obtained at par for each 20 ordinary shares subscribed by them. The remaining 50 founders shares were offered for public subscription, on the same terms.

Hammond and Co. had been retained as Electrical Engineers for the construction of the works, and produced an extremely optimistic report. 'Gentlemen', they wrote:

> 'In accordance with your instructions we have made a careful calculation of the cost of the proposed Central Station for Leeds, and we estimate that a Station containing three 100,000 watt plants (for night working) and one 50,000 watt plant (for day working) with all necessary mains laid in the scheduled streets, convertors, meters, etc., and suitable buildings and foundations (but exclusive of land) will cost £37,500. We think there would be no difficulty in completing the works by the first of November next, if put in hand forthwith, in order to be in time for the coming winter's lighting.
>
> The capacity of the above named works is, in our opinion, such as to permit of the wiring for 20,000 eight-candle power incandescent lamps, which, on the basis of 11/– per lamp per annum (the actual average elsewhere under similar conditions) would bring in a gross revenue to the company, when the plant is fully employed, of £11,000.
>
> We estimate the cost of working, repairs and depreciation, maintenance, coal, etc., for the above named output, at £5,574, leaving at the disposal of the Directors the sum of £5,426 p.a.'

Further evidence of the probable profitability of the new company was furnished by the inclusion in the prospectus of the financial results of five electricity companies, which showed dividends ranging from 3 per cent for the Keswick Company to 10 per cent for the St. James' and Pall Mall.

The attractions of the prospectus proved to be sufficient. In the first three days 109 people took out 7,296 shares, which provided enough initial working capital to enable the company to begin its operations.[14] The shares were not fully-paid-up at once, of course; 10/– per share was required on application and £1 more on allotment, with a further call of £1 on 1 June, so that the first capital in hand was nearly £11,000, with a further £7,296 due in another three months. This enabled the company to fulfil its first requirement, which was to deposit £5,000 with the Board of Trade under the terms of the Purchase Order. This was a show of earnest; the money would gradually be repaid as money was spent by the company on new plant and works.

Work begins

Advertisements for land brought many replies and from these a plot of about an acre was chosen in Whitehall Road at the side of the River Aire, no more than a few hundred years from the centre of town. Although more expensive than anticipated, the site promised future economies; the river would provide free water and cheap coal transport, and the closeness to the potential customers meant a big saving in cables and did away with the need for a separate office in town. The company were assured, too, of a moderate income in the form of rental from the firms occupying the portions of land not required, and there was also ample room for expansion if needed in the future.[15]

The intention (as expressed in Hammond's letter) was to complete the station and works by November 1892, and as far as the Press were concerned, there seemed little difficulty in meeting this target. As late as 7 September, the *Yorkshire Post* reported that 'construction of the works and plant is proceeding apace and is expected to be ready for November, as anticipated.' On the 16 September, the *Electrician* stated that 'mains are now being laid – supply is expected in November.'

There was also good news from the Board of Trade, who on 18 October wrote confirming their technical approval of the company's system, which was described thus:

> '. . . a high pressure alternating current supply at constant pressure to transformers fixed at first in suitable positions on consumers' premises, but when a sufficient density of supply has been secured, the distribution will be carried out by means of low pressure distributing mains fed from converting stations.
>
> The distributing mains consist of continuously insulated cables laid in cast-iron conduits, under the footways where possible.'

The approval was subject to acceptance of certain regulations and conditions which the Board of Trade had made which were 'for securing the safety of the public and for ensuring a proper and sufficient supply of electrical energy.' These were in two parts. The first thirty-nine regulations (safety) defined such things as the minimum insulation resistance, and the maximum current allowable in circuits, ensuring in total that the system was technically capable of giving a safe supply. The second group of regulations (supply) sought to protect the customer in such ways as requiring the pressure of supply to be declared and within a specified variation.

Public response to the company in this first few months was encouraging. Applications for lamps were being received in increasing numbers (8,000 lamps had been applied for by the end of December)[16] and in September supply contracts began to be signed.[17]

Good progress was being made, too, in the laying of the cast-iron cable ducts in the streets. This was a job being done by staff of the Corporation Highways Committee, who had laid 10,000 yards by the end of December.[18]

It's a pity, then, that Yorkshire House-to-House were actually in no position to give a supply of electricity, especially in November, as expected. The building of the station had become a problem, particularly the site foundations where great difficulties were encountered due to the treacherous nature of the soil near the river.[19] The weather that autumn was unkind, too. Heavy rain at the end of October caused severe flooding along much of the River Aire, from Kirkstall in the west to Hunslet in the east, and cannot have helped the contractors in their work.[20]

It was soon apparent to the company that they would not be able to give a supply to their new customers, and therefore not be able to fulfil their contractual obligations. They were faced with an embarrassing Christmas failure, with all the attendant bad publicity that that would bring. There was no alternative but to hire a portable steam engine which was installed in a temporary hut and used to drive one of the dynamos. This was connected to the new system and (according to the *Electrician* of 5 May 1893) supply was first made available on 14 December 1892.

As it turned out this was a wise decision, for the next day there occurred an accident at the Whitehall Road site which was to delay even more the completion of the permanent station. Exact details are not clear as the local newspapers all had slightly different stories. The *Yorkshire Post* reported that a floor collapsed; the *Leeds Times* claimed that a girder was being hoisted to the roof when the rope broke.[21] However, the *Leeds Evening Express* (15 December) seems to have the most comprehensive and convincing report of the accident which apparently occurred in this way:

> 'For some time a large engine shed has been in the course of erection, and some days ago the principals for the support of the roof were placed in their positions on the top of the building. The props were still up, and men had been engaged working on that part of the building all day. This afternoon two of the great girders, or principals, fell with a loud crash; one of them bringing with it a portion of the brickwork on which it had rested. Three of the men who were working in the vicinity have had marvellous escapes from being killed, but have escaped with bruises more or less severe.'

Two men went to the Infirmary with severe bruising; a third was seriously shaken but did not need treatment. The *Leeds Evening Express* was of the opinion that 'the accident will delay the erection of the premises somewhat considerably,' a conclusion which the months were to prove extremely perceptive. The station was not ready to supply electricity until the first day of May 1893.

Notes and references

1. Prospectus in *The Times*, 21 February 1888.
2. Jackdaw, 16 February 1889.
3. *Electrician*, 19 June 1891.
4. *Ibid*, 9 May 1890.
5. *Ibid*, 19 June 1891.
6. *Ibid*, 1 April 1892.
7. *Ibid*, 1 February 1889.

8. *Yorkshire Post*, 20 September 1890.
9. *Electrician*, 1 April 1892 and 15 February 1895.
10. *Ibid*, 16 August 1889.
11. *Yorkshire Post*, 26 March 1892.
12. *Ibid*, 26 March 1892.
13. *Electrician*, 19 February 1892.
14. *Leeds Mercury*, 29 December 1892.
15. *Yorkshire Post*, 29 December 1892.
16. *Ibid*, 29 December 1892.
17. All details of contracts and connections are taken from 'Records of Mains and Details of Service Connections', Yorkshire House-to-House Co.
18. *Yorkshire Post*, 19 December 1892.
19. *Ibid*, 29 December 1892.
20. *Leeds Mercury*, 22 October 1892.
21. *Leeds Times*, 17 December 1892.

Chapter 11

The opening of the new premises

The official opening

Despite the fact that the temporary plant ran without a hitch, giving a supply of great steadiness and regularity, the Yorkshire House-to-House Company must have felt great relief when the engineers completed their exhaustive trials and handed over the new permanent station in Whitehall Road. Although regular supply started on 1 May 1893, the official opening by the Lord Mayor of Leeds – Alderman John Ward – took place on 10 May, before a large crowd of important personages.[1]

There were representatives from many towns and cities, from as far afield as Cardiff, Southampton, Dover, Wigan, Glasgow, and Edinburgh, and more locally from York, Halifax, Wakefield, Sheffield and Bradford. As some dignitaries were making their four hour express train journey from London, though, many high functionaries of Leeds were making the opposite journey in connection with the Leeds Consolidation Bill, so there was a pronounced lack of local gentry at the day's celebrations.

Robert Hammond spent some time in the afternoon in his shirt sleeves, explaining the intricacies of electricity supply to inquisitive visitors, but at 5 o'clock, upon the arrival of the Mayor and Mayoress, he donned his frock-coat and tall hat to join the crowd at the ceremony.

To start the proceedings the Chairman of the Board of Directors, Mr. Grosvenor Talbot, introduced the Mayor, who received a cordial reception. After a short speech in which he expressed sorrow that he was not opening the Corporation's Electricity Works, he went into the switchroom to formally turn on the light.

Unfortunately, during the Mayor's speech someone had turned off the steam to the engines, so that when he closed the switch, nothing happened! This slight defect was soon put right though, and the Mayor once more closed the switch. Amidst great applause, light sprang up in every direction.

There then followed a tour of the works which took until about half past six – time for dinner at the Queens Hotel. About 200 guests were present and were treated to a first rate dinner, excellent wines and a number of speeches, all short, crisp and witty. The representative of *Lightning* magazine was very impressed with the whole affair: 'If a successful start be an augury for the success of the Company,' he wrote, 'then the Yorkshire House-to-House Company should be very successful indeed.'

The new premises and plant were described by various publications in a nearly identical form, as though taken from a publicity hand-out provided by the company. This particular description, though not unique, is condensed from the *Electrical Review* of 19 May 1893:

> 'The buildings are divided into engine-house, boiler-house, switchroom, test-room, workshop and offices. The engine-room is large and lofty, and 68 feet long by 63 feet wide.
>
> The plant installed . . . consists of two 200 i.h.p. engines driving two 100kW alternators and one 100 i.h.p. engine driving a 50kW alternator. The plant is capable of supplying 7,500 35 watt lamps. The engines are of the horizontal coupled compound condensing type. The speed of the larger engines is 90 revolutions, the ordinary working pressure is 125 lb. per square inch, but they will work without difficulty up to 140 lb., and give each 250 i.h.p. as maximum load. This provides for a good margin sufficient to meet all possible contingencies. A noticeable feature is the strength of the crankshaft, and the length and surface of the bearings.
>
> The engines are so arranged that they can be worked non-condensing in case of any failure of condensing water.
>
> The details of the smaller engine are as follows: it has a speed 120 revolutions, steam pressure is 125 lb. per sq in, with a maximum of 140 lb; the fly-wheel is 10 feet in diameter. The fly-wheels of the large engines are 14 feet in diameter. The fly-wheels are each grooved for rope driving, ten ropes of 1 and 1/4 inches diameter being employed.
>
> The alternators are Lowrie–Parker multipolar type, similar to those in use at West Brompton, Madrid, Dublin, Eastbourne, and Brighton. It is claimed that in the dynamos at Leeds there are many points of improvement over earlier machines. The working pressure is 2,000 volts, and, as in the case of the engines, allows a good working margin. In order to provide for the perfectly safe working of the plant at every point, the terminals to which the high tension mains are connected, are securely boxed in by polished wood lagging, so that they cannot be touched except by the proper official.
>
> The exciters for the alternators were driven by ropes from a grooved pulley on the alternator shift. A special feature with all these alternators is that the grooved pulleys are not allowed, as in the earlier machines, to overhang, but are supported by outside bearings, which give a solidity and ease in working, which amply repays the slight extra cost on the first outlay, and the after cost of lubrication. The bearings are all long, and have large surfaces, being provided moreover, with a water jacket and water fittings for use on emergencies.

THE BOILER HOUSE

The boilers are of the Lancashire type, and are three in number. They are 30 feet long by 8 feet in diameter; they are specially designed to work at a pressure of 140 lb. per square inch. Bennis's mechanical stokers are fitted to the boilers, these stokers are worked by means of a small engine driving by a belt on to a shafting and then on to the shafting of the stokers in the usual way. A Green's economiser is used, consisting of 192 tubes.

In case the engines are worked non-condensing, a feed-water heater is provided, by means of which the exhaust steam is used to heat the feed water.

Fig. 24. *Lowrie–Hall–Parker alternator, used by House-to-House (*Electrician*, 7 November 1890)*

Although a good supply of water was obtained by a river site, the character of the water is found to be very bad, there being a great deal of suspended matter in it. In order to meet this difficulty two water filters have been supplied each capable of filtering not less than 2,000 gallons per hour.

The main steam pipes are connected up in the form of a ring, thus providing two ways by which the engines can be supplied with steam, viz. from either end. The diameter of the steam pipe is 10 inches, and the whole ring is carried on brackets fixed to the wall of the engine-house. The whole of the steam pipes were duly tested by hydraulic pressure to 300 lbs per square inch in the station.

THE SWITCHING APPARATUS

The switching apparatus is placed in a special switch-room, and certainly makes a most impressive show. The apparatus is constructed under the Lowrie–Hall patents, and is very much on the lines of those in use at West Brompton. The apparatus provides for the simple working of each dynamo on to any circuit.

A synchronising board is provided, which is necessary when the alternators are run in parallel or when the load is changed over from one alternator to the other. The plant, however, is designed so that it can be run continuously in parallel.

While a great deal of discussion has gone on on the subject of running alternators in parallel, it is, we believe, the claim of the company's engineers that only

Fig. 25. *Lowrie–Hall transformer (*Electrician, *7 November 1890)*

in the works laid down by them in England is it the regular practice to run in parallel, by which means a great economy in work is effected.

THE GAY PARALLELING DEVICE

The main feature of parallel working to which objection can be taken is the risk of total failure of the supply in the event of an accident occurring to any of the units which are run on the bus bars. When (this) happens the disabled machine practically short-circuits its neighbour through its armature coils. The fuses of the others of the group are already loaded with the outgoing supply current, and therefore the excess due to the fault at once causes them to blow one after the other, thus completely closing down the station for a time. Mr. Gay has devised a method of arranging the fuses between the dynamos and bus bars as a whole and the various sections of the bus bars, that, in the event of such an accident taking place, the auxiliary fuses go before the main cut-outs could act, and instead of the latter being affected they are left to perform their legitimate duty of protecting the individual machines from an excess of current. The auxiliary fuses form the connecting link between parallel and non-parallel running, and

Fig. 26. *Lowrie—Hall transformer (*Electrician, *7 November 1890)*

when they act the machines are left to continue the supply, running non-parallel.

The switchboard is complete with all the necessary instruments for showing current or pressure, for switching exciters on or off and for regulating the voltage, so as to preserve a constant potential at the lamp terminals of the consumer, in accordance with the special regulation on this head made by the Board of Trade.

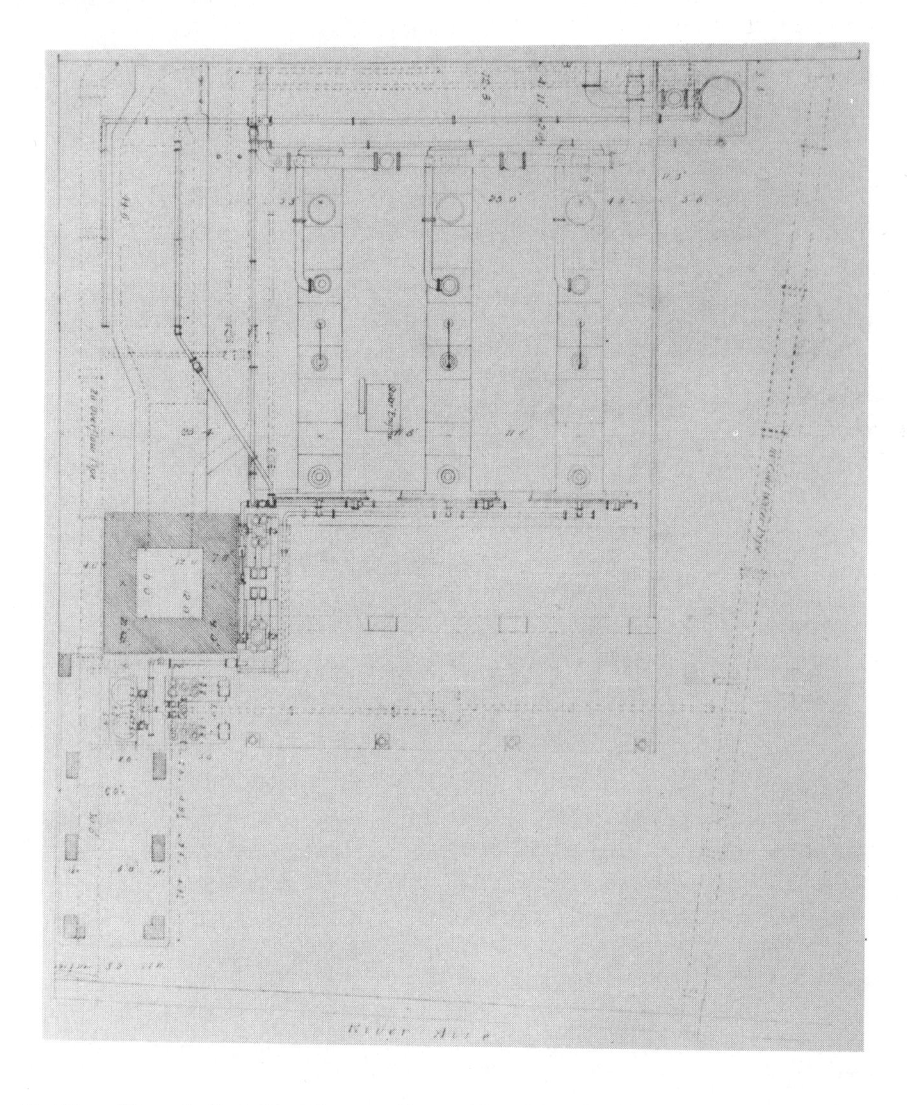

Fig. 27. *Plan of Yorkshire House-to-House Electricity Supply Station – boiler house (Yorkshire Electricity Board archives)*

THE SYSTEM OF MAINS

The Lowrie–Hall system of distribution has been employed, the particulars of which may be stated as follows:

The distribution of electricity generated on the Lowrie–Hall system is effected by means of copper standard cables, insulated with vulcanised India-rubber and laid in a specially-designed system of pipes combined with service and junction boxes.

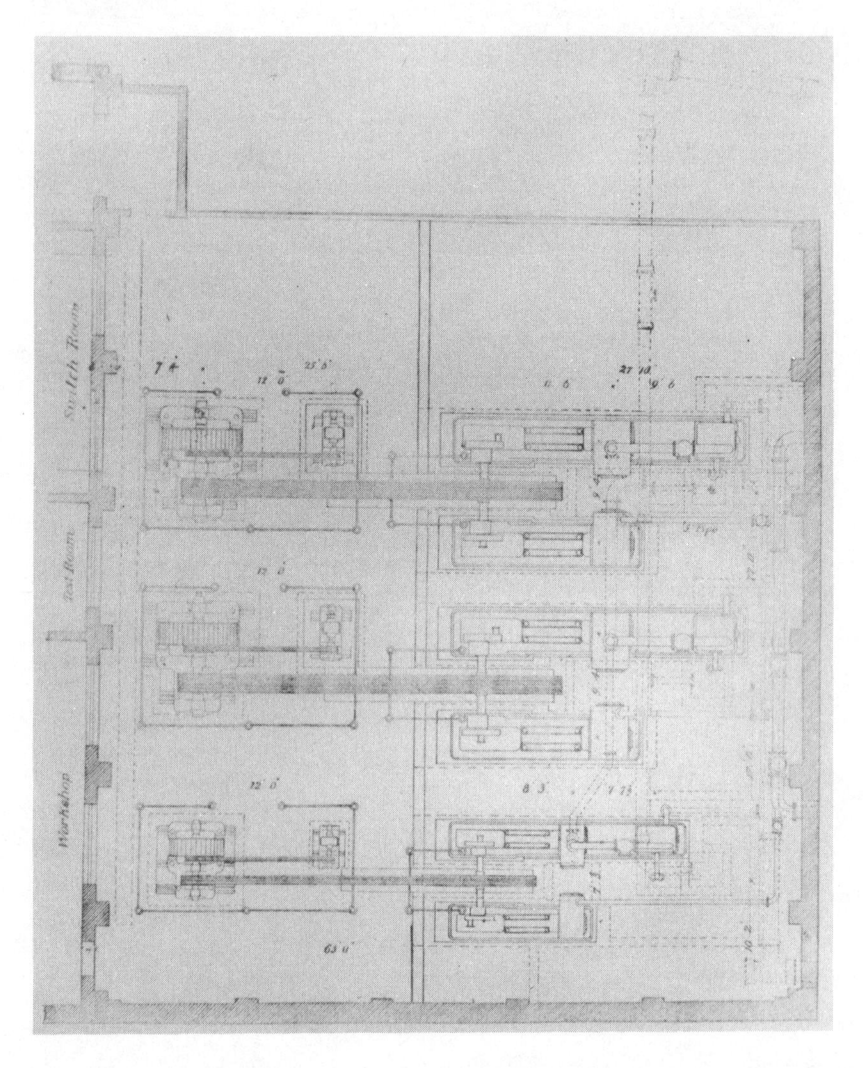

Fig. 28. *Plan of Yorkshire House-to-House Electricity Supply Station – dynamo room (Yorkshire Electricity Board archives)*

Each junction box has a loose cover, held in position by a cast-iron clamp. The culvert connection from the box to the consumer is made by a wrought-iron pipe. The surface boxes are in all cases placed at the corner of each road, and in the case of a long straight run they would be placed at a distance of about 80 to 100 yards apart. The surface boxes are made in two parts, the bottom part being permanent and the upper part with the cover being adjustable, so as to allow its surface to be raised or lowered as the road level may vary, the lid or cover being filled in with wood, macadam, or pavement according to the construction of the road. From the surface box the cable can be drawn in or out where required. Where passing round corners the cable is placed on a revolving drum, which entirely prevents the possibility of abrasion. The largest cable required for the work specified can be laid in half-mile lengths. The jointing of branch cables for consumers' houses is done by cutting away part of the insulation of the main cable, which consists of India-rubber, in the junction or surface boxes, and soldering the copper conductor to the conductor of the branch cables; the joint is then lapped over with rubber the full thickness of the original insulation and vulcanised on the spot, and the cable well-taped and mechanically protected, which makes the insulation of the joint practically equal to that of the cable itself.[2]

The cables of the India-Rubber, Gutta-Percha and Telegraph Works Company are 7/18s, 7/16s, 19/16s, 19/14s, all insulated with pure rubber, then vulcanised rubber of best quality, taped, and the whole vulcanised together, then braided and coated with preservative compound known as Silvertown Class L, they are then finally braided over all a second time and covered with preservative compound. The minimum insulation per mile of the cables is 5,000 megohm, after 24 hours immersion in water at 60°F and one minute's electrification.

THE CONVERTERS

The converters for reducing the pressure to domestic purposes are of the Lowrie–Hall type, are placed on the consumers' premises in special chambers, or in iron cases. The high pressure wires come direct on to one side of the converter, and the low pressure is taken direct from the other side through the secondary fuses and the meter into the lamps. The fuses for the moderate size transformers are placed in glass tubes, which are fixed on the sides of the transformers. In the large transformers fuses are placed in porcelain troughs, which are fitted on to a stout wooden frame fixed to the side of the transformer.

The consumers' current is measured by Shallenberger meters.[3]

TESTS, PRICE PER UNIT, ETC.

The following is a resumé of the tests:

Coal slack consumption 2.45 lb. per i.h.p. (equal to 1.71 of Welsh steam coal).

Water evaporated 8.6 lb. per 1 lb. of coal.

Water evaporated per i.h.p. 21.11 lb.

Electrical efficiency of dynamos 93 per cent.

Under the provisional order the rate of charge authorised is 8d per Board of Trade unit.

The scale of discounts is as follows:

For not less than 250 hr use (per quarter) of the maximum supply demanded 10 per cent.
For not less than 375 hr use (per quarter) of the maximum supply demanded 15 per cent.
For not less than 500 hr use (per quarter) of the maximum supply demanded 20 per cent.
For not less than 600 hr use (per quarter) of the maximum supply demanded 22½ per cent.
For not less than 700 hr use (per quarter) of the maximum supply demanded 25 per cent.

In addition to the discounts accruing under the above tables, a discount of 5 per cent will be allowed on the net amount of accounts paid within one month from the date rendering, and 2½ per cent on accounts paid within two months.

The effect of the discounts above mentioned is to reduce the price of 8d per unit to about the following net charges:

hours use (per quarter)	*price per unit*
250	6 3/4d
375	6½d
500	6d
600	5 3/4d
700	5½d

The whole of the plant has been erected under the supervision of Mr. Robert Hammond, MIEE, the consulting electrical engineer of the company, and the total cost, excluding land, has been about £27,000. Considering the output of the station in effective kilowatts, this seems a rather high sum, and appears to be somewhat in excess of what many other high tension stations have cost. The cost per kilowatt works out at £108.

The plans for the buildings were prepared by Messrs. Milnes and France, Architects, Bradford, and the works were given out as follows:

BUILDINGS – Messrs. Gould and Stephenson for mason and joiners work; Messrs. J. and H. Smith for iron work; Mr. Joseph Lindley, Mr. H. Season, Messrs. T. Cordingley and Sons and Mr. E. W. Walker, for plumbers, slaters, plasterers, and painters work, respectively.

GENERATING PLANT – including engines, boilers, dynamos, switchboard, converters, and all accessories, Messrs. John Fowler and Co. (Leeds) Ltd.

METERS – Messrs. The Westinghouse Electric Co. Ltd.

CAST IRON PIPES AND BOXES FOR STREET CULVERTS – The Stanton Ironworks Company.

INDIA-RUBBER COVERED CABLES – The India-Rubber and Gutta Percha Co. Ltd.'

The station was finally ready. Admittedly, it had been expensive, but this was because the directors, full of confidence, had built a station much larger than their immediate needs. Although the installed capacity to begin with was only 250kW (enough for about 7,500 lamps) expansion would be cheap and easy.

Who could afford the new light?

The question now was, would expansion be necessary; would anyone want the electric light? Indeed, could anyone afford the electric light? For let there be no

Fig. 29. *Yorkshire House-to-House – lid of street box, outside Municipal Buildings, 1984.*

mistake, the new illuminant was expensive, and beyond the means of many.

To illustrate this point let us consider first the small houses of the poor – the 'two-up, two-down' terrace houses typical of so much of the city. The four rooms of these houses, if lit by electricity, would probably need an installation of only six lamps of 35 watts each, giving a total load of 210 watts. Assuming that the company's discount system was based on realistic and probable hours of usage, the minimum requirement in a quarter would be for 250 hours, for a discount of 10 per cent giving a preliminary account of £1.15s.0d and a final bill of £1.13s.3d if payment was prompt enough to earn the extra 5 per cent discount.

To the people who lived in these houses 33/– was a large sum of money, a lot more than they could afford when they earned only 21/7, as tram conductors did for a working week of more than 60 hours[4] – and that was before deduction of income tax at 7d in the pound. Yorkshire House-to-House was unlikely to be successful in such areas.

But what about the larger houses in the more prosperous suburbs. In Brompton we have already seen that the average installation was 60 lamps – a house with this many lamps was without doubt the home of one of the wealthy middle classes, probably with servants. A fully discounted bill for 250 hours use in a quarter would be of the order of £15, probably less than a servant's wage for a year and not beyond the means of the householder. And apart from the ability to pay, there would doubtless have been pride and status to consider, a Victorian 'keeping up with the Jones' in the matter of household gadgetry. Similar households in Leeds would surely be the potential market for Yorkshire House-to-House.

As far as commerce and industry were concerned there seemed little doubt that the electric light would be popular despite the extra cost, provided this was not excessive. To start with there was an increasing realisation that a lot of money could be saved on decorations because of the cleanliness of electricity. For example, the Savoy Theatre had required decorating every year when lighted by gas, whereas it had gone eight years without decoration since the installation of elcectricity.[5]

Then there was the brightness of the electric light to consider, which not only improved productivity in offices and factories, but was also beneficial to shopkeepers who found that they had a bright and impressive way of showing off their wares. An attractive electrical display could actually increase custom, as Brooke Bond later found when they opened their refurbished premises in Boar Lane. The *Leeds Express* of 12 April 1894 tells us that the illuminated window display was so eye-catching that 'large numbers of people congregated round the premises to have a glimpse of the powerful illumination.' There seemed little doubt that investment in the electric light would produce dividends for the man of business.

The significance of these various eeconomic and practical arguments was obviously not lost on the directors of House-to-House, for the new system was concentrated on the commercial and shopping centre of the city and on the areas wherein lay the large and prosperous households of the rich and successful, such as Park Square and Headingley.

However, none of the system had been installed south of the River Aire in the townships of the poor, like Hunslet and Holbeck, and as this was where most of the Leeds industry had grown (Hunslet was long famous for its steam engine) there was no immediate prospect of the connection of industry to the new system. Apart from this omission, it did appear that the directors had made a promising start, but the first few months would test the correctness of their decisions.

Notes and references

1. Description of the opening is from the *Lightning* magazine, 18 May 1893.
2. At the time of the opening there were 12,000 yards of culverts and the furthest lamp was three miles away, (*Electrician*, 5 and 12 May 1893).
3. O. B. Shallenberger was the Chief Electrician of Westinghouse, America, and died in 1898 aged 38 years, (*Electrician* 11 February 1898).
4. *Yorkshire Post*, 3 January 1895.
5. *Leeds Mercury*, 7 March 1894.

Chapter 12

The first connections[1]

Despite the reliability and success of the temporary generator, the restriction on supply caused by the lack of a proper station was a blow to Yorkshire House-to-House, lasting as it did for virtually the whole of the company's first winter lighting season. There was also a restriction on supply caused by the fact that not all the network was ready as anticipated in the preliminary areas. As a result many of the lamps for which contracts had been signed could not be connected at once, and it is perhaps ironic that the very first customer to sign a contract with the company had to wait several months for a supply despite the fact that he was also a director of the company. This was Arthur Lupton, who signed his contract for 27 lamps for the premises of the family firm at the corner of Wellington Street and Aire Street on 9 September 1892. It was four months later, on 17 January 1893, that his supply was connected.

By the end of 1892 another forty-six contracts had been signed and a total of 2,922 lamps were contracted for – an average of 64 lamps per contract. However, half of these customers had to wait until the new station was ready before they could receive a supply, and two customers had to wait until November 1893 for their connections.

More fortunate were the proprietors of five shops in Boar Lane and Commercial Street which were connected to the new system within a few weeks of contract and were the first premises to receive a supply when the temporary network began operation on 14 December. The first two shops to be connected (on 9 December) were in Boar Lane; C. J. Hardy (outfitters) was obviously a large shop as it had an installation of 349 lamps, while Pickles Confectioners were rather small, with an installation of 32 lamps. Pearce and Sons (watchmakers) in Commercial Street with 90 lamps was connected on 12 December, the fish shop of Timothy Newby (Boar Lane) had 72 lamps connected on 13 December and on the day the network was made live, Smith J. Wales (tailors) in Commercial Street were connected with 35 lamps. These five pioneers provided a total opening load of 578 lamps.

By the end of December another six premises had been connected and the load had grown to 995 lamps. In the next four months there were connections to another nineteen premises so that at the end of April – just before the permanent generating station opened – there were thirty customers with a total of 2,063 lamps, or about 82kW. This was only about a third of the lamps which had been contracted for, as there were still fifty-two contracts outstanding for another 4,305 lamps.

The availability of the new generating station at the beginning of May saw a large increase in the number of connections as the company was at last able to catch up on its contractual obligations. On the first day of May, thirteen premises were connected to the system, ten more were connected on the second day and there were twenty-five connections made over the rest of the month. These included 725 lamps in the premises of the *Yorkshire Post* (on 9 May) and 66 lamps at Spring Bank on 15 May. In this one month connections rose from 2,063 to 5,582 lamps, an amazing 271 per cent increase.

This growth, of course, could not be sustained for the rest of the year but nevertheless an average of more than 600 lamps was connected each month to the end of the year. At the end of 1893 there was a connected load of 11,634 lamps (or about 370 kW) supplied to 139 customers who had used 130,068 units of electricity.[2] The Company had now managed to deal with nearly all the outstanding contracts and only five customers were waiting for a supply.

Problems of load

The response over the first twelve months was much greater than the company could ever have hoped for, and although this must have been a source of great relief and satisfaction to them it was, too, a source of embarrassment, for they had a continual struggle to supply the load applied for. The main difficulty was that the company were under the handicap of not knowing what load would result from the lamps which were to be connected to the system; all they could be sure of was that the lamps would never be switched on all together, for one of the characteristics of an electricity supply system is that the maximum load demanded from the system is always less than the total installed load, the proportion of maximum demand to installed load being the 'diversity'. The Leeds system being new, the directors were not sure what the diversity would be and were forced to treat the connections with caution, for fear of overloading the dynamos. In fact the temporary dynamo – of only 50kW capacity – at the end of April was supplying a total installed load of 82kW, so that the diversity (assuming the dynamo to be fully loaded) was 50 dividend by 82, or 0.6.

The dynamo was not fully loaded all the time, of course, for the 'load factor'[3] in Leeds at that time was only about 14 per cent, which meant in effect that lamps were lit only for about three or four hours each day. In one way this low utilisation (today it's more than 50 per cent) was of some benefit to Yorkshire House-to-House, in that they were easily able to make cables dead during the day without inconvenience to consumers, thus facilitating the work of extending the high voltage mains.

Tha main effect of low utilisation, however, was to impose a severe financial burden on the company, for it meant that assets were only bringing in income for 14 per cent of their lives, and were thereby preventing a lowering of unit production costs. The importance of this effect was demonstrated by Mr. Hammond, who proved that typical unit costs of electricity supplied could be reduced from 2½d to ld if the load factor could be increased from 12 per cent to 50 per cent. It was for this reason that discounts were offered by electricity companies for extensive consumption of electricity.

Although utilisation was low and was to remain low, the increasing applications for supply soon convinced Yorkshire House-to-House that an extension of their plant was necessary. As early as the beginning of July 1893 – only two months after the opening of the permanent station – the local papers were reporting a planned extension of 40 per cent.[4] The proposal was to install another engine of 200 i.h.p. and a corresponding dynamo of 100kW; despite the size of the extension the increase in capital expenditure was expected to be only 10 per cent as the works had been built to allow easy additions.[5]

The new dynamo was ready and running in November,[6] although of course not required all the time. In fact, according to the *Leeds Mercury* of 5 March 1894: 'two machines are constantly running, and being coupled in parallel, it is impossible for them to vary in speed more than one 10,000th of a revolution. A third machine is brought into use at busy times, and the fourth is kept as a standby.'

But this was not enough! By the end of 1893 it was apparent that the diversity was much different from the 0.6 originally estimated and in fact the company Chairman, Mr. Grosvenor Talbot, reported in March 1894 that: 'through some unexplainable reason, three-fourths of their lamps were going at one time, a far larger proportion than any other electric company that he knew of.'[7]

Plans for a new plant

This realisation caused the company considerable problems, in that for a short time before Christmas it was compelled to decline further applications, 'the producing power being fully taxed' during the time of heaviest load.[8] This was a serious matter, for the company had plans to extend the district of supply and also wished to connect more houses on existing routes; if the newly extended plant could not supply the existing load it would certainly be inadequate for the load anticipated for the following winter. The directors had no hesitation therefore in announcing a huge increase in generating plant, to be installed as soon as possible, which would more than double capacity from 350kW to 750kW. This new plant would be in operation (it was hoped) at the beginning of the next winter period.[9] The Directors also thought that, if the load justified it, an agreement could be made with Fowlers to 'trade-in' the small 100 h.p. engine and dynamo and replace these with a 400 h.p. engine and 200kW generator.[10]

This rapid expansion of the business must have been something of a pleasant surprise to the directors, but at the same time it placed a great strain on their ability to raise the necessary capital. At the end of 1894 capital expenditure had reached £64,000. But where had it all come from?[11]

The first source – and the most important – was by share issue. The 100 founders shares had been taken fully-paid-up (at £5) almost immediately. Of the 10,000 ordinary shares on offer, 7,366 had been taken up by the start of July 1893 and various calls were made on these shares until they also were fully-paid-up, at the end of the year.[12] The directors held a fifth of these shares between them.[13]

To finance the first extension the directors offered the remaining 2,634 shares, giving preference to existing shareholders. These were soon taken up,[14] but only small calls were made on these shares because of the increasing calls on the earlier shares, and this second issue was not fully-paid-up for another eighteen months.[15]

A second source of capital was a mortgage of £6,000 which the company obtained on the land it owned at Whitehall Road. Two-thirds of the land was still let off to private businesses, the income of £142 p.a. going a long way towards the annual mortgage interest payment of £240.[16]

When the second extension was planned, the directors felt that increased calls on the existing shares issued would finance most of the proposed work, the shortfall being only in the region of £8,000. They felt it was more to the advantage of the company that the money should be borrowed from the bankers rather than go to the shareholders and the public and ask them for more capital. They believed this step should wait until a much larger amount was required. Accordingly an overdraft was arranged, secured by a £10,000 mortgage debenture.[17]

Further difficulties

Urgent provision of new working capital was not the only problem to occupy the directors, however, especially in the first few months of working, when it must have seemed that fate was conspiring against the company. The first difficulty of 1893 was a coal strike, which could have been disastrous for the new company, but for the foresight of the directors, who were able to build up adequate coal stocks before the stoppage took place. As it was, the station was able to continue generating, although at a greater expense than normal.[18]

Later in the year, sales were seriously affected by one of the brightest autumns and winters on record, which meant a great reduction in the hours of lighting. A director of one company in London declared that up to Christmas they had lost £1,000 in consequence of the bright weather, and generally it was estimated that £80,000 had been lost on that account.[19] The new company was already at a disadvantage that first year in that the first four months – normally months of long lighting hours – had been unproductive as the permanent station had not opened. During the temporary generation period the 6,828 units which had been sold had merely covered expenses, and the hire, erection and removal of the temporary engine had cost £73.

Under the circumstances then it was something of a surprise that a net profit of £1,155.6s.1d was announed at the end of December 1893. As all the plant was still under its twelve month guarantee the directors decided that there was no requirement to make a depreciation allowance and resolved to dedicate £1,000 of the profit to reducing the preliminary expenses (such as the management expenses during the construction of the works) which had amounted to £2,186. Although necessarily paid for at the time out of capital, these expenses did not represent increased assets and the directors felt obliged to pay them off from the revenue account.[20] The balance of £155 was carried forward.

All things considered, the directors must have been very pleased with the start made by Yorkshire House-to-House, and indeed in admitting this the Chairman, Grosvenor Talbot (at the company's first Annual Meeting on 6 March 1894) could only wonder 'that the company should have been allowed to have the opportunity of such a valuable opening as they now possessed of supplying the light to Leeds. He hoped they would avail themselves of that opportunity by doing their best for the citizens.'[21]

As an earnest of their intent in this respect the Chairman then confirmed that prices were to be reduced by 12½ per cent on 1 April, the maximum charge thus being reduced from 8d to 7d a unit. 1893 had been a surprisingly good year for Yorkshire House-to-House and the directors were determined that 1894 would be even better!

Notes and references

1. Details of contracts, connections and numbers of lamps connected are taken from an analysis of 'Records of Mains and Details of Service Connections', Yorkshire House-to-House Company.
2. *Lightning*, 15 March 1894.
3. Load factor is the ratio of the actual units generated to the product of the maximum demand of the year and the total hours of the year. In a rapidly growing system it is not easy to determine the true maximum for there is a gradual increase of load throughout the year, culminating in a maximum demand at the end of the year. It was usual to use the year end maximum, despite the artificially low load factor this produced, but a truer figure could have been obtained using an average of the maximum demands at the start and end of the year. In Leeds the load factors for 1896 were 13.44 or 16.34, using year end and average maximum demands respectively. These arguments and the resultant figures were presented by R. Hammond in a paper read for the IEE on 24 March 1898, printed in the *Electrician*, 1 April 1898.
4. For example, *Yorkshire Post, Leeds Express, Yorkshire Evening Post*, 3 July 1893.
5. *Leeds Mercury*, 5 July 1893.
6. *Electrical Engineer,* 2 March 1894.
7. *Leeds Express*, 6 March 1894.
8. *Leeds Mercury*, 28 September 1894.
9. *Ibid*, 7 March 1894.
10. *Yorkshire Post*, 26 January 1895.
11. *Electrical Engineer*, 1 February 1895.
12. *Lightning*, 15 March 1894.

13. *Yorkshire Post*, 3 July 1893.
14. *Leeds Mercury*, 7 March 1894.
15. *Electricial Engineer*, 1 February 1895.
16. Letter from Mr. Green, Secretary of Yorkshire House-to-House, to *Electrical Review*, 2 March 1894.
17. *Electrical Engineer*, 1 February 1895.
18. *Leeds Mercury*, 7 March 1894.
19. *Ibid*, 7 March 1894.
20. *Yorkshire Post*, 7 March 1894.
21. *Leeds Mercury*, 7 March 1894.

Chapter 13

Respectability at last

The year 1894 began slowly for Yorkshire House-to-House for even after seven months the number of connected lamps had only increased by 10 per cent. But then potential customers obviously began to think of the forthcoming dark autumn and winter evenings and connections began to increase rapidly; at the end of the year there were 19,858 lamps connected to the system, an increase of nearly 90 per cent over the 12 months.[1]

At the annual meeting at the end of the year the Chairman was able to report on another successful year. Profit was a healthy £3,672 which enabled the company to write off the preliminary expenses and set aside £500 to a reserve fund. This still left enough to pay a dividend of 4 per cent, which was an important publicity exercise for a company hoping that expansion would require more public investment before too long.[2] To emphasize the success of the company, the Chairman pointed out that of the other twenty-one electric light companies in the country, the seven corporations paid average dividends of only 2 3/4 per cent, while the dividends of the leading private companies were no more than 4 per cent.[3]

A successful start

There was little doubt that Yorkshire House-to-House had made a very successful start to their enterprise, and that the electric light was now thoroughly respectable and acceptable. There were reports that all the major buildings being erected in Leeds were dependent entirely upon the electric light as an illuminant. Among the more important buildings already lighted by the company were the Leeds Parish Church, the Great Synagogue, the Yorkshire Post Offices, the Leeds new School of Medicine, portions of the General Infirmary, of the School Board Offices, and of the Philosophical Hall, the Leeds Library, the Co-operative Society's new premises in Albion Street, the Law Institute, the Leeds Club, the Conservative and Unionist Clubs, the Royal Exchange Club, the Builder's Club, the Bank of England, the Yorkshire Penny Bank and a considerable number of other banks, insurance buildings and offices, as well as hotels, shops and private residences. Several persons and firms who had previously obtained electric light from private installations had now adopted the company's supply.[4]

The Great Northern Railway Co. was one of the customers, too, Yorkshire House-to-House giving a supply at 2000V to a sub-station in the goods yards at Wellington Street. From the sub-station a cable system, two miles in length, supplied thirty Brockie–Pell arc lamps and five incandescent lamps at low voltage in the various yards, warehouses and granaries.[5]

Not that there was always complete confidence in the supply provided by Yorkshire House-to-House. The *Leeds Times* queried whether the electric light ought to be absolutely replied upon by the Church wardens of the Parish Church, who had made no arrangement in their scheme for the retention of small gas jets for emergency use.[6] 'Without them,' the *Leeds Times* suggested, 'a panic in the dark might easily result in serious accident. And, by the way, every one of the outer doors of the Parish Church open from the inside, and the failure of the dynamo, plunging the church in darkness, would in all probability lead to a mad rush for those doors, which of course would be jammed. Nobody wants to see the Parish Church a scene of shrieking and fainting women, climbing over each other in frantic panic.'

This pessimistic piece was printed in September 1894 and funnily enough the doors of the Parish Church were tested only two and a half years later. There was what the *Yorkshire Post* called 'an awkward incident' when the splendid final chorus of Dvorak's 'Stabat Mater' – the highlight of the Holy Week service in April 1897 – was interrupted by the sudden failure of the electric light. Fortunately there was no confusion, no panic; the Benediction was at once pronounced and the congregation 'dispersed in the most orderly fashion possible.'[7] Yorkshire House-to-House were quick to point out in the press that the failure was not due to any defect in their system,[8] but the Vicar pressed ahead with an arrangement of gas lights for emergency use.[9]

More extensions

Also in September 1894 the respectable Leeds newspapers were describing the extensions to the works at Whitehall Road which were then nearing completion.[10] At least most of the works were nearing completion, for there had not yet been agreement with Fowlers to change the small engine and dynamo. However, two new 400 h.p. engines and their associated 200kW dynamos were nearly ready and three new Lancashire boilers (to provide for the future Fowlers replacement) of 140 lb. pressure were already installed and connected to the steam ring system.

Although similar in design to the old engines, the new ones were much bigger; their weight was double, at 18 tons, and the fly wheels were each 17 feet in diameter and grooved for fifteen ropes, while the wheels of the older engines were but 14 feet diameter and used only ten ropes. The bedplates of these monsters were supported on 12 feet of ashlar stone set on 3 feet of concrete. To accommodate the new plant the engine room had been almost doubled in length (to 124 feet) and the boiler house was similar.

The extensions, when finished, provided the company with a total generating capacity of 750kW worked by engines of approximately 1500 h.p. and this was how the plant stood through the winter at the end of 1894. The distribution system consisted of eight miles of cables covering much of the city centre, as far eastward as the Parish Church, westward to Woodhouse Square, north to the top of New Briggate and bounded in the south by the river, except for a short main just over Leeds Bridge. There was also the long cable up to Spring Bank and its neighbours in Headingley. There were plans for immediate extension from Woodhouse Square up the prosperous Clarendon Road,[11] and from Briggate up into Chapeltown which 'in view of the shops and prosperous public houses that lined that thoroughfare, could not but prove successful.'[12] A ground for satisfaction at the year end was the fact that the company had never had a breakdown, believed to be a unique record.[13]

In an attempt to encourage a greater utilisation of the system the company offered electricity for purposes other than lighting at a rate of 3d per unit. The expectation was that this load (for cooking or power) would be used mainly during the day.[14] There is little evidence to suggest widespread use of this tariff in 1894 but there is a report (*Yorkshire Post*, 28 September) that the Leeds and County Unionist Club had used an electric oven since the opening of the club, and it was 'singularly effective'. The same report noted that the electricity company were themselves using a 6 h.p. motor for driving the scrapers of the fuel economiser and mechanical stokers.

At the end of 1894 an announcement was made that the rate for units used for lighting was to be reduced from 7d to 6d on 1 July 1895, with certain modifications and simplifications to the discount system. The chairman explained that the company had already lighted large works and shops and large houses in Headingley, 'but it must be remembered that capitalists would pay for what might be considered luxuries, and that the great thing for a company to consider was how likewise to reach small shops and houses of a less extensive character.'[15] There was no doubting the sense of purpose of the directors. Yorkshire House-to-House was going to be successful.

Notes and references

1. *Electrical Engineer,* 1 February 1895.
2. *Ibid*, 1 February 1895.
3. *Yorkshire Post*, 6 February 1895.
4. *Ibid*, 28 September 1894.
5. *Ibid*, 30 December 1893.
6. *Leeds Times*, 15 September 1894.
7. *Yorkshire Post* 13 April 1897.
8. *Ibid*, 15 April and *Leeds Mercury*, 16 April 1897.

9. *Leeds Mercury*, 23 April 1897.
10. *Yorkshire Post* and *Leeds Mercury*, 28 September 1894.
11. *Leeds Express*, 5 February 1895.
12. *Yorkshire Post*, 6 February 1895.
13. *Leeds Express*, 5 February 1895.
14. *Electrical Engineer*, 1 February 1895.
15. *Yorkshire Post*, 6 February 1895.

Chapter 14

Municipal acceptance

Although the first two years of business had been very good for Yorkshire House-to-House, with a continued growth that must have exceeded all their expectations, 1895 turned out to be the year which finally established the position of the company beyond doubt; this was the year in which they received the accolade of municipal acceptance, with all the added status that this brought.

The vital weekend was 17 and 18 March, as explained in this extract from the *Leeds Mercury* on the following Monday:

> 'The electrical current by which the Leeds Town Hall and Municipal Buildings are lighted is now supplied by the Yorkshire House-to-House Electric Lighting Company. At 9.30 on Saturday evening the work commenced of connecting the lights in the two buildings with the mains of the Electric Lighting Company, and this was completed about seven o'clock on Sunday evening. Afterwards the current was turned on, and tried in various rooms, in all of which it proved very satisfactory.'

The load thus supplied by Yorkshire House-to-House in the Municipal Buildings was 5,961 lights, an increase of almost a third on the company's system, and was supplied via eight converters and seventeen meters.

The 'most fatuous blunder'

Not that the Council's decision to allow private supply to public buildings had been made quickly or easily, for this was not the Council's way. In fact argument ebbed and flowed for four years, starting only three years after the new electrical plant had been installed in the Corporation's premises in 1888. It only took this short period of time for the deficiencies of the plant to become evident. The *Yorkshire Post* commented on 12 May 1891 on the numerous complaints which had been made about the electric light and reported that the Sub-Corporate Property Committee were considering asking the Borough Engineer to prepare plans for a fresh installation.

No action seems to have taken place as a result of this committee's deliberations, for on 12 December the *Yorkshire Post* was again commenting on the electric light, which it now described as 'fitful'. One of the immediate reasons was said to be the unsuitability of the coal supplied for the generating station, but in fact the whole system was badly made and badly sited. The engines and dynamos, for example, were in a completely unfit position in the basement of the Municipal Buildings where the temperature in which the attendants worked was often in the region of 90°F.

The wiring, too, was soon found to be defective, as much of it was too thin to carry the current with safety. The company who insured the buildings refused (in January 1892) to renew the insurance unless defects were put right, so the Corporate Property Committee recommended that the Council spend £500 on a new installation in the Free Library and a complete overhaul elsewhere.[1] They also recognised the necessity of finding new and larger premises for the generating plant and once again cast their eyes on the shed at the rear of the Fire Brigade Station in Park Street.[2]

At the Council Meeting in March 1892 there were anxious and recriminatory voices raised at the proposed expenditure.[3] Ald. Sir E. Gaunt claimed that the system was without proper fuses, such fuses as there were being mounted on wood and not slate, and the wires and lamps were too small. The whole system needed re-modelling, he alleged, and much more than £500 was required.

Mr. Winn denounced in strong terms 'the waste of public money which had taken place in an installation which was now acknowledged to be absolutely ineffective.' Mr. Wilson described the whole business as 'the greatest muddle the Corporation had ever go into.' According to him the boilers were altogether inadequate, both engines had broken down at one time, there was no arrangement for economising steam or for feeding the boilers properly, the dynamos had been out of order and the chimney was so bad that even their own smoke inspector had threatened to summon the Committee. The whole thing was an absolute 'fizzle'. Ald. Spark did his best to refute the allegations of negligence pointing out that the equipment had been the best available at the time but that vast improvements had been made since.

Despite the criticisms, the Council agreed to spend the £500, although recognising that this would not eliminate all the problems. And indeed this was so, for in January the following year the *Yorkshire Post* was complaining bitterly of the 'daily discharge of volumes of dense smoke from the chimney' which had reached proportions serious enough for the Sanitary Committee to complain to the Corporate Property Committee.[4] Several Councillors asked that the whole lot be removed immediately, supply to be given by Yorkshire House-to-House, but the company were then still running the temporary generator and were unwilling to commit themselves.[5]

Again, in October 1893, the *Yorkshire Post* thundered out about the 'most fatuous blunder' and the 'Radical ineptitude', claiming that many parts of the Municipal Offices had been in semi-darkness. 'The expensiveness of the scheme', it declaimed, 'has far surpassed the worst predictions of the most ardent opponents of the project.'[6]

Matters were brought to a head at the start of 1894 when Yorkshire House-to-House, confidence boosted by the success of their first eight months operation, offered to supply the Town Hall and Municipal Buildings for a price of 4d per unit.[7] Not wishing to act with unaccustomed haste the Corporate Property Committee asked its Chairman to have a look with the City Engineer at the options available to them, and in June a sub-committee was appointed to help them in their deliberations. In September their report was submitted to the Electric Lighting Committee, a report which investigated three courses of action:[8]

1. To continue generating at the Municipal Buildings.
   The cost over the last year for the 123,646 units used had been £3,429.6s.6d, or 6.70 pence per unit. However, urgent work was now required to install a large new chimney, to provide a standby engine, and to prevent smells and vibration in the buildings. This would involve a large capital outlay.
2. To accept the company's offer.
   The cost of lamp renewals and labour would probably increase the cost from 4d to 6.57d per unit.
3. To install a new generating station approximately two miles from the Town Hall at the refuse destruction plant at Kidacre Street.
   Cost for this scheme would be about 7.21d per unit, high because of the large capital investment needed to set up a new station and lay the necessary mains, and despite the free steam available.

After an hour's discussion the ELC unanimously decided to recommend acceptance of supply by Yorkshire House-to-House, but for a trial period of only two years.

In the light of Council decisions which had been taken earlier in the year, this recommendation – with its lack of capital investment – was inevitable. On 4 July the Council had agreed to apply to the Local Government Board for a Provisional Order empowering them to raise a loan of £100,000 for municipal purposes as 'the borrowing powers on this account were now absolutely exhausted.' In fact according to the Mayor they were 'slightly in excess of being exhausted.' During discussion it was revealed that a 2/– rate would have to be levied to meet the expenditure. A scandalised Ald. Scarr said that 'this Council was the most extravagant Council he had ever known, and they were getting worse, 'a point of view not unpopular among Councillors. Who now would suggest capital expenditure on another Electric Light Experiment?[9]

When the full Council discussed the ELC recommendation on 3 October the result was a foregone conclusion, especially when the state of the plant was detailed. The tin chimney was now full of holes and in imminent danger of collapse, the engines were running at twice the speed they were designed for and the electricial plant worked at a pressure so high that an accident was always likely. On top of this the offensive smells and noise were 'an abominable nuisance' to both employees and public in the offices and art gallery.

Sorrow was expressed by one Councillor that the Yorkshire House-to-House Central Station was not under Corporation control but he was soon cut to size by Ald. Scarr, who said that they were not fit to manage it, in fact they were fit to manage very little. 'There was not one of them who managed their own business as they managed things here,' he retorted. The resolution was adopted with only one vote against.[10]

However, old habits die hard. Two months later there had been no action, and no improvement. On 14 December the *Yorkshire Post* revealed that the lights were so poor in the Library that the magazines, newspapers and books could not be read. Furthermore, the chimney was now so bad that draught in the boilers had almost disappeared and gallons of oil were having to be poured onto the fires to keep them going. There was a definite air of frustration in the newspaper's reference to 'the difficulties that inevitably come in the way of the Leeds Corporation when they have set their hands to a thing.'

Disagreement over terms of supply

The difficulty in this case was that the two parties could not agree terms of supply, the Corporation objecting particularly to the company's request for a minimum annual equivalent to the purchase of 120,000 units. The objection was so strong that on 26 December the Press[11] reported that negotiations were at an end, and on New Year's Day 1895 the Corporate Property Committee met to consider laying down new plant at the Municipal Offices![12]

There is no doubt that harsh words from various sources were secretly conveyed to that meeting, for the committee ignored its agenda and instead authorised two of its members to re-open negotiations with the company, who, it was understood, might be willing to modify their previous proposal.[13] And so it proved, for on 4 January 1895 agreement was at last reached and a contract was signed.[14]

What concessions were made and by whom is not revealed, nor is the reason for the delay in connection. On 5 January supply was expected to be in two weeks' time, but as we have seen the actual changeover date was on the 17 and 18 March.

This was not a moment too soon. Apart from the appalling condition of the machinery, the atmospheric conditions in the offices above had become so bad that at least a third of the thirty or forty clerks in the Rate Office had been on the sick list in recent weeks.[15] It was a great relief to the officials in the Municipal Buildings to be able to get on with their work in a clean and healthy atmosphere, and without the continual 'steamboat-like thud' that had been a source of annoyance over the years.[16]

Heat was still expected to be a problem in the basement because of the presence of the eight converters which gave out considerable heat while performing their duty of transforming the electricity to a low voltage, but an electric fan had been installed to prevent the heat from becoming a nuisance.

On 21 March 1895, only a week after connection of the new supplies, the

Yorkshire Post published what must have been the only letter ever written in praise of the ELC This was from 'One who was there' who congratulated the ELC on the brilliancy and steadiness of the electric light in the Victoria Hall compared to the variable and sombre light that all had been accustomed to for so long.

Notes and references

1. *Yorkshire Post*, 22 January 1892.
2. *Ibid*, 18 February 1892.
3. *Ibid*, 3 March 1892.
4. *Ibid*, 21 January 1893.
5. *Electrician*, 14 April 1893.
6. *Yorkshire Post*, 27 October 1893.
7. *Leeds Daily News*, 6 March 1894.
8. *Yorkshire Post*, 14 September 1894.
9. *Ibid*, 5 July 1894.
10. *Leeds Mercury*, 4 October 1894.
11. *Yorkshire Post* and *Leeds Mercury*, 26 December 1894.
12. *Leeds Mercury*, 2 January 1895.
13. *Yorkshire Post*, 2 January 1895.
14. *Ibid*, 5 January 1895.
15. *Ibid*, 19 March 1895.
16. *Leeds Mercury*, 19 March 1895.

Chapter 15

Trams and lamps

The success of Yorkshire House-to-House in winning the contract for supplying the Town Hall with the electric light did not, unfortunately, signify the complete acceptance of private enterprise by the Municipal Authority, for there were two other electrical projects in hand by the Corporation which the Council had every intention of doing themselves. As it turned out, the second project was to depend upon the first, which was the extension of the electrical tramway system.

Electric trams

Electric trams had first appeared in Leeds in 1891 when the Thomson Houston Company of America had obtained agreement from the Corporation to construct and operate a short experimental line running between Harehills and Oakwood. The trains were garaged in premises in Stanley Road, near the junction of Beckett Street and Harehills Road (now a Refuse Department depot), and in the same premises a dynamo driven by a gas engine gave a DC supply of 500V on the overhead wires. It's strange to think that despite constant displays of hesitancy and incompetency on so many occasions, in this particular field the Council showed bravery and farsightedness to a great degree, for this short electrified line was the first practicable electric street tramway in the whole of Great Britain.[1] The first tram ran on 29 October 1891.

The electric trams quickly established themselves and despite opposition which occasionally bordered on the hysterical (mainly due to the alleged unsightliness of the overhead conductor system) the little line was at last recognised as a success. The Corporation therefore decided upon a large scale experiment hoping to be able eventually to superimpose the electric system over the whole of the City's existing tramway network. Accordingly, in early 1895, the Corporation promoted a Parliamentary Bill for the purpose of acquiring authority to extend the tramways and to construct stations for generating electrical power. Objections to the Bill were lodged by various Railway Companies and by the National Telephone Company, but these objections were soon overcome by explanation and negotiation. More seriously, Yorkshire House-to-House objected to the Bill because of the Corporation's intention of erecting their own central station, with the possibility of supply to Council premises and street lighting. The company were of the opinion that the Order under which they worked entitled them, and them alone, to offer such a service.[2]

Explanation and negotiation (between the company and the Leeds Parliamentary Sub-Committee) failed to produce agreement and Leeds Corporation were

Fig. 30. *Thomson–Houston tram on the Roundhay line, at the Roundhay terminus, Oakwood (author's collection)*

eventually successful in obtaining Parliamentary approval for their scheme despite the company's objections.

Early in 1896 the Council began planning the first line to be electrified, a stretch of more than six miles running from the Canal Gardens at Roundhay, through the city, to Kirkstall Abbey. Unable to provide electricity for the line by statute, the company (in May 1896) offered the Corporation a guaranteed supply of electricity at cheap rate in the hope of pricing a Council generating station out of the market.[3] The offer was a simple guarantee of constant and adequate power at a rate of 2.3d per unit based on an estimated requirement of 292,000 units per year, the contract to be for only three years. It was pointed out that this cost was inflated by the possibility of the company being left with spare equipment on its hands at the end of the contract, so an alternative cheaper suggestion was made. This was for the Corporation to provide their own dynamos which the company would install at the Whitehall Road station. The dynamos would then be available at the end of the contract for the Corporation to use elsewhere. The company explained that acceptance of either of these schemes would prevent foolish expenditure on plant which the Corporation might eventually find inappropriate when the six mile line was working and fully evaluated. The Corporation would then be free, when extending the electric tramway system, to adopt any improvements in plant that might then be available.

This offer was ignored completely by Leeds City Council who had already invited tenders for all the plant and equipment needed for the Council to set up their own system, and , in fact, the tender invitations persuaded Yorkshire House-to-House that they had underestimated the quantity of electricity which would be used each year. As a result, at the end of June 1896, they put in a new offer to the Council; based on a consumption of 584,000 units per year the company offered electricity at 1.75d per unit. They were even willing to reduce this to 1.25d per unit if the Council provided the dynamos. This latter rate would involve the Corporation in an annual payment of £3,041.13s.4d, a considerable economy, it was claimed, on the capital expenditure and annual charges otherwise involved.[4]

The company initiative was doomed from the start, however, for this was one project the Council were almost unanimous in wishing to undertake themselves. After considering the tenders, the Council voted overwhelmingly in July to reject the offer of Yorkshire House-to-House and to accept a cost estimated at almost £50,000 to set up the new line,[5] a cost which included about £20,000 for a new generating station.[6] There was very little support for the private company's offer, most opposition to the Council's decision being centred on the cost of the accepted tenders, which although mainly from Leeds firms[7] were not always the cheapest.[8] Councillor Lowdon, who had recently inspected the auto cars at the Imperial Institute, also claimed that electric traction would soon be made obsolescent by the oil–gas motor which he believed would ultimately be used very extensively for traction purposes.[9]

These minor objections did not deflect the Council from their purpose, oddly resolute as they were on this singular occasion, and in late 1897 the electric tramway system was working. The new generating station had been built in The Calls at the north end of Crown Point Bridge (the engine house is still there, although used for other purposes now) and sufficiently large to meet possible extensions of the system for many years. There was also a series of workshops and car sheds in Kirkstall Road and twenty-five tramcars were purchased, although thirteen of them were delayed because of an engineers' strike.

The cable installation was considerable, for cables had to be laid from the generating station to the extremities of the line at Roundhay and Kirkstall. The Telephone Company, too, had considerable cable work to do, for they had agreed to take down their overhead lines from the new electrified route and put their wires underground.[10]

In none of this work was the Yorkshire House-to-House Company involved; they would never be asked to provide supplies for traction. Another company with no future in Leeds traction was Thomson–Houston, who, unsuccessful in the tendering, collected together their equipment and departed the city.[11]

Street lighting

While this debate (and the development of electric trams) was taking place, the Council also found themselves in the midst of arguments on another electrical

Fig. 31. *A Leeds trolley bus in the first year of operation – at the Woodcock Pub, Farnley terminus, 1911 (author's collection)*

topic, that of street lighting. There had long been complaints about the gas lamps in the streets. The newspapers often grumbled that the lamps served only to 'render darkness visible'[12] and the *Yorkshire Evening Post* commented that 'strangers to Leeds find its paths very dark.'[13]

The dawn of a new era broke in January 1896 when the Council had erected a 'refuge' for pedestrians at the junction of Boar Lane, Wellington Street and Park Row. The traffic here was very great with an endless stream of lorries, tram cars, omnibuses and other vehicles, and to those no longer nimble the refuge was considered to be an 'inestimable boon'. At night the refuge was illuminated with a 30 c.p. illuminated lamp, the lamp and the power both supplied by Yorkshire House-to-House. This was probably the first public electric street lamp in Leeds.[14]

A year later there was a growing realisation that the poles supporting the tram wires would be ideal for the placing of street lamps, indeed in many places this was desirable for the new poles were being erected only a few feet from existing gas lamps. Not only was this unsightly but the pale gas light was seriously restricted and undesirable shadows were produced.[15] There was talk at first of transferring the gas lamps onto the new poles, which being hollow would easily allow the

Fig. 32. *The last horse bus on the Farnley route, 1911 (Leeds city Libraries)*

installation of gas pipes up the middle, but soon there were demands for the lamps to be lit electrically. After all, the poles were supporting electric wires for the trams and it should be easy to take a supply to the lamps. However, it wasn't as easy as that. The Tramways Committee pointed out that 'as the power of the electric current by which cars are to be propelled will vary very much at different points it would be too uneven in its operation to be partially utilised for lighting the streets.'[16] Separate supplies would have to be provided for street lighting.

The Lamp and Highways Committees were given the task of recommending what lamps to use and how to supply them; before long they had decided that the streets to be lit would be Wellington Street (from the Central Station), Briggate, from top to bottom, and the whole of Boar Lane. Twenty-nine 1000 c.p. arc lamps would be required, erected at 40 yard intervals and would be provided by the Corporation.[17]

Thus far decisions had been easy, but now it was necessary to decide who would

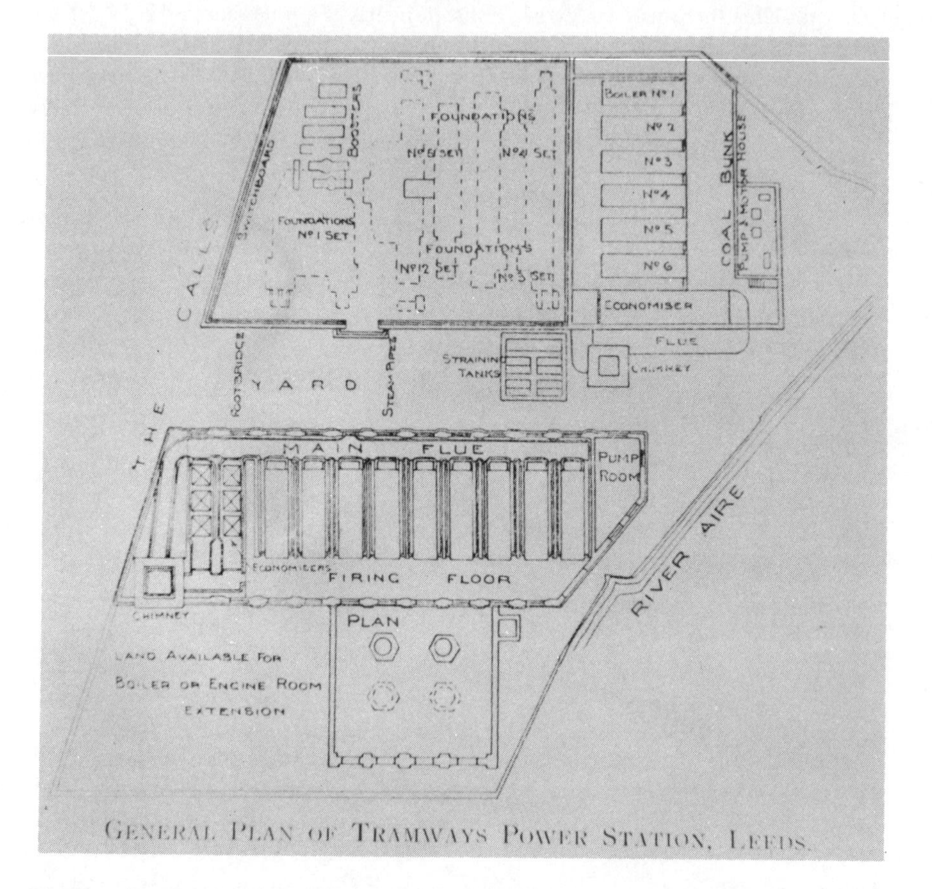

Fig. 33. *Plan of Leeds Tramways Electricity Station, Crown Point, 1906 (*IEE Electrical Handbook for Birmingham and Leeds 1906*)*

supply the current and for this there were two alternatives. First, the Corporation could supply electric lighting current from its station at Crown Point on separate lighting circuits. Dr. Hopkinson, who had been adviser and consultant to Leeds Corporation for the tramways project, was asked to investigate the practicalities of supply and its cost.[18] Second, the whole scheme could be passed over to Yorkshire House-to-House, who had already, at the end of 1896, put in a very attractive offer (they believed) to the Corporation. This was to supply an adequate and steady light over the prescribed area for a yearly sum of just under £500, less than the cost of gas and, according to the company, less than the Corporation would be able to achieve.

The Committees did not make immediate recommendations and at the end of March 1897 Yorkshire House-to-House were becoming alarmed that once again they were to be excluded from public supply; they had already been prevented from supplying the new tramways and now looked like being prevented from supplying the street lamps. They therefore circulated a letter among the members of the Lamp Committee in which they reiterated the economical advantage of their

Fig. 34. *Crown Point Works – assembling dynamos (Leeds City Libraries)*

proposals and indicated their strong desire to meet the Committee. They stated their willingness to make any reasonable modification in their proposals to meet the Committee's wishes, but reminded the Committee that they had applied for a Provisional Order and entered on the work of electric supply for Leeds at a time when the business was regarded as very speculative, being, to some extent, induced by the prospect of eventually being called upon to do street lighting. They concluded by telling the Committee that they were in a position to give an immediate supply of current for lamps in any of the main streets.[19]

This pleading was to no avail. The Council decided that the whole operation would be controlled by the Corporation and towards the end of May advertised for tenders for the supply and fixing of arc lamps. On the 2 June 1897 the Council accepted the tender of Mr. William Wharam, an electrical contractor of Clay Pit Lane,[20] at a cost of £988.10s.[21] This was almost double the £575 which had been allowed for[22] but to be fair to the Council and Mr. Wharam the requirements had been considerably extended, the power of the lights now being 2000 c.p. (not 1000 c.p.) and the numbers being increased from twenty-nine to forty.[23]

Completion of the installation was promised for 1 October but unfortunately a strike in the engineering trade prevented delivery of the arc lamps from the manufacturers[24] and it was not until Christmas that work finished and switch-on was possible. The Chairman of the Lamps Committee performed the official ceremony on Wednesday 22 December,[25] on a night when a dense fog hung over the city. The system proved to be a 'distinct and unqualified success', according to the *Leeds Daily News,*[26] although the gas lamps were left in position in case of a failure of the electric light.[27] They seemed very small and quite insignificant in comparison.

The *Leeds Daily News* reported that Briggate had never looked so gay, for besides the new street lamps the shopkeepers had also used the electric light to bright effect in their windows and the more enterprising had adopted the new form of electrical advertising. The best display of this type was undoubtedly that of Messrs. Pattinsons Ltd., the highland distillers of Leith, Ballindallock and London, who had secured the front of the upper storeys of Crampton's Hotel in Briggate. On this frontage the name of the firm and its product was set forth in gigantic letters, adorned by no less than 370 electric lamps in white and red. By an ingenious device the lights alternated at short intervals throughout the evening. A similar sign advertised the virtues of Sunlight soap and a third was erected by Rowntrees to promote that company's cocoa. This display of electric illumination was very new and very exciting, so that (according to the *Leeds Daily News*, again) 'a number of ladies and gentlemen, who were detained at the Coliseum for the annual performance of the "Messiah", were quite loud in their admiration of the new system.'

Notes and references

1. Eight electric tramways were already in existence in Britain, but without exception these were supplied with electricity in such a way that they were much too dangerous to run along public streets. In Portrush, for example, supply was from a live rail – a T iron – placed on standards 3 feet high at the side of the track: Ryde Pier had a similar system. The Newry/Bessbrook system was supplied from a centre live rail (with huge electricial losses in wet weather) and Blackpool had a live rail in a tube laid in the ground with a slit in the top. (*Tramways: their Construction and Working,* D. K. Clark, 1894).
   Meanwhile, in America, electric transport had been growing apace, and in 1890 there were 2,523 miles of electrically operated track and 5,592 electric cars within the United States and Canada, with a total investment of £7,166,000 (*Electric Railways and Tramways*, Philip Dawson, 1897). It was from this advanced technology that Leeds chose its equipment, provided and installed by the Thomson-Houston Company, which used the overhead conductor system, expensive to install but suitable for application throughout the town with safety.
   Twenty years later trams became so popular in Bradford as well as in Leeds, that there was considerable pressure from adjacent towns for the sytems to be extended beyond the

boundaries of the two cities. In some places, however, legal powers to lay tracks proved difficult to obtain, so the joint Tramway Committees had the novel idea of using 'trackless trams', which were of course trolley buses. In June 1911 Leeds opened a trolley bus route from Aire Street to New Farnley while by arrangement Bradford opened a route from that town at the same time. Leeds and Bradford were thus the joint originators of trolley buses.

2. *Leeds Mercury*, 17 May 1895.
3. *Yorkshire Post*, 20 May and 3 June 1896.
4. *Ibid*, 22 June 1896.
5. *Leeds Mercury*, 2 July 1896.
6. *Yorkshire Post*, 2 July 1896.
7. *Ibid*, 26 June 1896.
8. *Leeds Mercury* 2 July 1896.
9. *Yorkshire Post*, 2 July 1896.
10. *Ibid*, 19 October 1897.
11. *Leeds Mercury*, 15 July 1896.
12. *Ibid*, 8 April 1897.
13. *Yorkshire Evening Post*, 2 April 1897.
14. *Leeds Mercury*, 25 January 1896.
15. *Yorkshire Post*, 21 January 1897.
16. *Ibid*, 21 January 1897.
17. *Ibid*, 18 March 1897.
18. *Ibid*, 4 February 1897.
19. *Ibid*, 2 April 1897.
20. *Ibid*, 20 December 1897.
21. *Leeds Mercury*, 3 June 1897.
22. *Ibid*, 13 April 1897.
23. *Ibid*, 19 October 1897.
24. *Ibid*, 19 October 1897.
25. *Ibid*, 28 December 1897.
26. *Leeds Daily News*, 23 December 1897.
27. *Leeds Mercury*, 23 December 1897.

Chapter 16

Growth

When all things were taken into consideration, it was soon apparent that Yorkshire House-to-House had not done too well in its dealings with Leeds Corporation, despite the connection of the Town Hall and Municipal Buildings to the company's system. As the directors knew only too well there was little prospect of load growth in the Municipal Buildings whereas the prospects of supply to the new electric tramway and lighting systems were immense, for huge extensions to both systems were surely to be expected over the forthcoming years. Despite these disappointments, though, the company continued to prosper, with load connected, units supplied and profits all increasing very satisfactorily.

Yorkshire House-to-House sets records

From the end of 1894 load grew steadily from 19,585 lamps (or equivalent) connected to the system of 49,150 lamps at the end of 1897 and units supplied increased at a similar rate, from 291,113 per year at the start of this period to 833,280 units supplied during 1897.[1]

Profit increased at an even greater rate, almost quadrupling over the four-year period from the £3,672 of 1894 to £13,395 in 1897,[2] so that large amounts of money were available for transfer to both depreciation and reserve funds, and dividends were increased from 4 per cent in 1894 to 5 per cent the next year and 6 per cent in both the following years.[3]

One of the reasons for the financial success of the Company was the tight control exercised by the directors and works manager over costs – particularly works costs – a fact frequently commented on by the magazine *Lightning*. Indeed the company often featured at the head of the record costs tables in that magazine.

At the end of 1893 the works costs during what *Lightning* referred to as the 'pioneer period' of the company were seen to compare very favourably with other undertakings during their pioneer periods[4] and in its class (of 100,000 to 150,000 units sold) it was the cheapest in the country, with a works cost of 3.20d per unit sold.[5]

The following year the works costs of 2.17d set a new record for the new class (the company had sold nearly 300,000 units) and it was noted that coal costs of 0.584d per unit had been bettered by only one station of any class and that was by Bradford which had the advantage of supplying almost double the number of units.[6] The year 1895 was even better, for the company achieved record low total costs (for any size undertaking) of 2.05d and coal costs were reduced to another new record of 0.3d.[7]

The amount set aside for depreciation and reserve (4.7 per cent of the capital expended) was the greatest of any private company and was beaten only by three local authorities (including, again, Bradford).[8] Not that this was altogether due to the astuteness of the directors, for in the matter of the depreciation fund, they had very little control. It was the Government who advised on the financial life of capital assets (such as cables and generators) and it was the Government auditor who had the power to declare what amount should be added to the fund before he signed the accounts or certified the amount of the capital.[9] The reserve fund was somewhat different, though, and the directors of Yorkshire House-to-House were always keen to extend this fund. They were fully aware that they were in a new industry subject to sudden change and believed that 'there were possibilities in electrical science which possibly might render advisable expensive alterations.'[10] If their equipment became obsolete because of a dramatic improvement in technology, renewal would have to come from reserves rather than depreciation. They would 'more willingly entertain' replacements if they had in hand a considerable reserve.

In January 1897 *Lightning* magazine was again full of praise for the Leeds company when the results for 1896 were published.[11] By this time the magazine had abandoned its classifications based on output – rapid growth of the undertakings was causing too many changes in the classes – and merely divided the companies into two groups: those who provided public lighting (and thereby held a cost advantage) and those who didn't. Yorkshire House-to-House did not supply street lamps and so was placed in the second group where it established new records for both works and total costs, at 0.96d and 1.73d respectively. *Lightning* again complimented the company's Manager and Chief Engineer, Mr. Harold Dickinson, and told this little story to demonstrate the 'spirit of frugality' which he engendered at the Leeds works:

> 'The agent of a certain class of instrument was explaining to a works manager how a sample of what he was showing failed to give satisfaction at the Leeds works, and casually suggested as a cause that the attendant had unfairly dealt with it by "dipping a handful of waste in a bucketful of oil," and applying the waste to the instrument in question. After the agent had departed, a driver who had overheard the conversation went up to his chief: "That were a lie, sir, about them samples," he said; "them people at Leeds never bought a bucketful of oil or a fistful of waste all at wunst in their lives."'

This was a spirit which persisted and again the following year[12] *Lightning* was enthusing over the company's results; now the Leeds undertaking had established a new record for works costs of 0.78d per unit, the oil waste water and stores were the lowest in the country, and the coal costs of 0.25d were beaten only by Stafford (at 0.24d) and Nelson (0.15d). *Lightning* was obviously not far from the truth when it printed the opinion that: 'for capable management no electricity undertaking in the kingdom can claim priority over that of the Yorkshire House-to-House Company.'

Another capable representative of the management of the Leeds Company was the Secretary, Mr. W. T. Green, who regularly won fulsome approval from *Lightning* magazine for the prompt publication of the company's results. Under statute the annual statement of accounts up to 31 December each year had to be published on or before 25 March the next year; Mr. Green normally managed to have his accounts cleared and in the hands of the printers a bare fortnight into January. Yorkshire House-to-House were invariably the first undertakers to publish their figures, a considerable achievement considering the vast amount of analysis required under the law.[13]

It's understandable perhaps that a public who had learned to dislike private monopolistic companies in service industries had demanded a frankness in the operation of this most modern industry but the degree of openness required was somewhat surprising. Indeed Robert Hammond – in a paper read before the IEE on 24 March 1898 – offered the opinion that the 'existence of such data is unique in the history of English industries, if not in that of the industries of the world.'[14] 'I fancy,' he added, 'that the Chairman of the Gas Light and Coke Company would shrink from such a rigid analysis of his cost sheets as that set out in the form prescribed by the Board of Trade for electricity undertakings.'

Fortunately there was no need for the Chairman of Yorkshire House-to-House to shrink from such an analysis for he could have had nothing but pride in the figures provided from the analysis; costs had started low and had decreased as the load grew. In fact, the only real worry facing the chairman and directors was how to cater for this load growth, which always seemed to be at a rate greater than they had anticipated.

There had been continued expansion at the works from the opening day, as we have already seen, until in the summer of 1895 the buildings were filled to capacity, Fowlers having completed the removal of the 50 kW generator and engine, and replaced them with new equipment rated at 200 kW.[15] The total works capacity was then 900 kW.[16]

New projects

But again this was not enough! In July 1895 the directors announced that they were considering the installation of 600 kW of new plant in buildings which would be built on that part of their site not yet developed. In consideration of this – and because the mains extension to Chapeltown was imminent[17] – the directors also announced an increase in capital, to be raised by the issue of the 9,900 unallotted ordinary £5 shares. The conditions of issue were 10/– upon application, and 10/– on allottment, with a guarantee of no further calls before January 1896.[18] The shares were offered at par to existing shareholders only, who were 'privileged to apply for any number not exceeding their present holdings.'[19] Purchase at par was indeed a privilege, for the shares had been steadily increasing in their market value. In the *Leeds Mercury* of 6 July, for example, £5 fully-paid-up shares were quoted at £6½ to £7 on the Provincial List and before long the new shares – £1 paid up – were being quote at 1⅜ premium.[20]

As the old year drew to a close, two projects were seen to have progressed with varying degrees of success. The extension of the mains to Chapeltown was almost finished – at a low cost of about £6,000[21] – although not, apparently, to everybody's complete satisfaction. A letter from 'L.K' in the *Leeds Mercury* of 17 January 1896 complained of the poor state of Chapeltown Road, accounted for by 'the careless way the flags have been relaid since the laying of the House-to-House Electric Light Company's lines.' Reinstatement has always been a problem!

Progress on the works extensions had been somewhat slower although the Chairman did announce in February at the company's annual meeting that orders had been placed for two 500 h.p. vertical engines designed to run at the high speed of 355 revolutions, and two Ferranti 300 kW alternators to be directly coupled to the engines. This was confirmed in *Lightning* a couple of months later[22] when it was revealed that the suppliers of the equipment would be G. E. Bellis and Co., Birmingham. The boilers and steam pipes were to be installed by Hicks, Hargreaves and Co. of Bolton.

It is difficult for us to realise what a stir this announcement caused, but there must have been considerable agitation in the City of Leeds, particularly in the Hunslet engine works of Fowlers, who had always supplied engines to both Yorkshire House-to-House and its London parent. Indeed Fowlers had always considered themselves to be major contributors to the success of Yorkshire House-to-House, advertising proudly and largely in *Lightning*[23] that the various low-cost records held by the Leeds electricity undertaking had been achieved by the use of their slow-speed engines and rope drives. Now, alas, they would have no part to play in the future of electricity in Leeds. The irresistable conclusion was that the requirements of power plant for electricity generation were growing apart from the requirements of agricultural power, which still was content with slow speed engines – Fowlers could not or would not keep pace with this new technology.

It was soon apparent, moreover, that Fowlers' loss would be much greater than had first been realised, for when the details of the contract were announced there was also a statement from the board declaring that arrangements were in hand to buy Brittania Mills,[24] a well let property of 5,700 square yards adjacent to the works.[25] There was no immediate prospect of development of the site but it would be an invaluable asset when the present site was no longer adequate. There was every prospect of large orders for new plant within the next few years.

Good news and bad news concerning progress

At the beginning of 1897, twelve months after the announcements, there was good news and bad news concerning progress. The good news was that the purchase of Brittania Mills had been completed successfully, at a cost of £22,819.[26] The bad news was that a six month strike in the building trade had prevented completion of the 1000 h.p. extension, which had been expected in September, and the company had found themselves short of generating capacity during the winter peak. To their great dismay this shortage had necessitated a refusal to connect load during the winter and the company had even found themselves discouraging applications, in an attempt to restrict the load! Although the units sold in the year were a new 'high', sales had been a lot less than expected. Without doubt the delay had been a setback for the company.[27]

There must have been considerable doubt as to whether the new plant would even be ready for the next winter, for it was not until the end of October 1897 that Yorkshire House-to-House announced the commissioning of their new plant[28] (which had been 'greatly delayed by labour troubles').[29] As if to celebrate their relief at the long-awaited availability of the generators the company also announced a reduction in charges: from 1 January 1898 units expended in lighting would be reduced in price from 6d to 5d and units used for power and heating would cost 2d, down a penny from 3d. There would of course be the usual discounts for increased usage and prompt payment. A not unimportant concession at the same time was the abolition of the charge hitherto made for connecting consumers' premises to the mains where the distance of the premises from the road did not exceed ten feet.[30]

These reductions in charges served only to increase the growth in unit sales and connections, as was intended of course, and expansion was so rapid that no sooner was the new plant in operation than a new extension of a further 900 kW was at once proposed.[31] By April 1898 two-thirds of this machinery had been installed and was operating[32] and the remaining 300 kW was commissioned during the summer. At the end of 1898 the capacity of the works had reached 2400 kW.

During the period of this extension of the plant and works, similar extensions to the distribution system were taking place. New mains were planned for Kirkstall and Hunslet[33] and in May 1897 the first reports appeared of underground transforming sub-stations, although the reports were perhaps not in the form the company would have preferred, occurring in a lengthy reference to complaints from shopkeepers about the appalling state of the streets. It was in the *Yorkshire Post* of 4 May that the complaints were reported, the irate and aggrieved traders of Briggate having held a meeting the previous day. There were emotional outbursts at this meeting about the disturbance to the pavements, where tramway supply cables were being laid, the road, where tram lines were being laid, and the large hole being dug by the electricity company. What was particularly upsetting was the fact that most of the work had been going on for more than four months, since Christman, and there seemed little prospect of completion.

Mr. Hannam, Chairman of the Corporation's Highways Committee, defended the Council's actions as best he could but stated that the 'transformer' pit, which was being dug out, was not his responsibility but that of Yorkshire House-to-House, in the same way as the one 'which had attracted attention near the Yorkshire Post offices.'

(A few months later, incidentally, Mr. Hannam again came under severe attack during electioneering for the fact that, despite his position on the Highways Committee, his own company actually obtained a great deal of civil engineering work in Leeds street works, giving rise to severe doubts about his integrity.)[34]

By August 1897 four underground sub-stations had been built – two in Albion Street, at the bottom and at the corner of Bond Street, and two in Briggate at the junctions with Boar Lane and Commercial Street.[35] These sub-stations were of great benefit to Yorkshire House-to-House who were now able to supply premises at low voltage and could dispense with the transformers previously required at each point of supply. These had been expensive to install and run, and incurred high electrical losses. It was much easier to make new connections to a 200 V system than to a 2000 V system, which had to be made dead for each connection, thus interrupting supply to consumers already connected.

The low voltage cables were of a design new to the company and had been specially approved by the Board of Trade. They had paper insulated concentric conducting cores with an outer lead sheath, embedded in asphalt composition, enclosed in tarred wooden troughing and further protected by bricks laid over the top.[36]

At the end of 1896, with all this work in prospect, the directors had recognised the need to increase capital in some way to pay for the proposed extensions. There were to be no half measures this time, they decreed, for the future looked rosy. They recommended to the shareholders a doubling of capital, a recommendation accepted at the Annual Meeting in January 1897.[37] To double the issue, each shareholder was offered new shares at par equal in number to his existing holding. In the next twelve months 18,948 shares were offered: 18,626 were allotted.[38]

To maintain the increasing momentum of growth the directors of the company issued an amazing circular, in March 1898. The autumn and winter of each year was always busy, they stated, as new customers were persuaded by the shortening days to accept the electric light and this put the company's workmen under undue pressure at this time of year. To overcome this the directors offered three months of free electricity to those whose premises were connected in the summer.[39]

'Such open-handed liberality almost makes one blush,' commented the *Leeds Express*, 'to think of the hard things that have been said at times against companies and dividend hunters. How unjust we sometimes are.'

'And yet,' pondered their correspondent (who admitted to being 'quite a child in such matters'), 'is the offer altogether as disinterested as it seems? Is there nothing else lurking behind? I wonder whether a considerable increase of new customers and trade would tend to increase the purchase money the Corporation must pay for the undertaking?'

And indeed, the Corporation had at last decided to buy out the Yorkshire House-to-House Company and to undertake the supply of electricity themselves in the City of Leeds. The prophecy of Jackdaw, so many years ago, had at last come true, for the takeover was going to be expensive.

Notes and references

1. *Electricial Review*, 15 February 1895 and 21 January 1898.
2. *Lightning*, 31 January 1895 and 27 January 1898.
3. *Electrical Review*, 15 February 1895, 31 January 1896, 22 January 1897, 21 January 1898.
4. *Lightning*, 15 March 1894.
5. *Ibid*, 31 January 1895.
6. *Ibid*, 31 January 1895.
7. *Ibid*, 30 January 1896.
   N.B. 'Works costs' were the costs attributable to purchase of coal, oil, waste, water and engine room stores; salaries of engineers and wages; and repairs and maintenance of buildings, engines, cables, etc. 'Total costs' were works costs plus rent, rates and taxes; and management expenses (administration, insurance, advertising, etc.).
8. *Lightning*, 30 January 1896.
9. *Leeds Mercury*, 3 February 1897.
10. *Ibid*, 5 February 1896.
11. *Lightning*, 28 January 1897.
12. *Ibid*, 27 January 1898.
13. *Ibid*, 28 January 1897, 27 January 1898.
14. Lecture printed in *Electrician*, 1 April 1898.
15. *Leeds Mercury*, 7 June 1895.
16. *Lightning*, 18 July 1895 – note that with the engines used the capacity of the dynamos were always considered to be half the h.p. of the engines, i.e. 1,800 h.p. was equivalent to 900 kW. The new high speed engines proposed in July 1895, however, were capable of driving 600 kW of dynamos per 1,000 h.p.
17. *Leeds Express*, 15 July 1895.
18. *Leeds Mercury*, 15 July 1895.
19. *Lightning*, 18 July 1895.
20. Provincial List, *Leeds Mercury*, 28 November 1895.
21. *Leeds Mercury*, 5 February 1896.
22. *Lightning*, 23 April 1896.
23. *Lightning*, 18 February 1897.
24. *Leeds Mercury*, 5 February 1896.
25. *Ibid*, 20 January 1897.
26. *Ibid*, 3 February 1897.
27. *Yorkshire Post*, 3 February 1897.
28. *Leeds Mercury* and others, 28 October 1897.
29. *Lightning* 11 November 1897.
30. *Leeds Daily News*, 28 October 1897.
31. *Electrician*, 21 January 1898.
32. *Ibid*, 8 April 1898.
33. To Meadow Lane, *Electrician*, 21 January 1898.
34. *Leeds Mercury*, 20 October 1897.
35. *Yorkshire Post*, 21 August 1897.

36. *Ibid*, 21 August 1897.
37. *Ibid*, 3 February 1897. It was intended to offer stock for £94,740, offering shareholders exactly one share for each share held. There were at that time, therefore, 18,948 shares already issued.
38. *Electrician*, 21 January 1898.
39. *Leeds Express*, 25 March 1898.

Chapter 17

Takeover

To such astute men of business as the directors of the Yorkshire House-to-House Company it must have been apparent at the start of their enterprise that there was little prospect of a long and independent life for their company, indeed there seemed to be only two possible results of their actions. It was clear that the undertaking would either fail – a possibility which had frightened away the Council in the first place – or it would succeed, in which case the Council would buy them out, for the Council couldn't allow an important public service to be in private hands.

It must have been no surprise, then, to read in the Leeds press on 28 August 1897 that two resolutions had been included on the agenda paper of the forthcoming meeting of the Council. The first called for the purchase by the Council of the Yorkshire House-to-House undertaking, as allowed for in the company's Order, the second proposed that negotiations be delegated to the Highways Committee.

Certain sections of the Press were delighted that these resolutions had at last been put forward. The *Leeds Mercury*[1] received the news 'with much satisfaction', for the Leeds Liberal Association had given formal approval of the resolutions the previous evening, and the *Leeds Express*[2] found the ideas 'gratifying' for it had often advocated the takeover and thus approved of the 'wise proposal'. The *Yorkshire Post* was more guarded, conceding only that the resolutions were of an 'important character.'[3]

A few days later details began to emerge of the cost to the Corporation of the takeover and the benefits which would accrue to the shareholders. The papers reminded the public that under the terms of the agreement, if purchase occurred within the first ten years, the Council were obliged to provide sufficient Corporation stock to ensure a 5 per cent annuity on the capital expended, plus further stock to provide a net dividend of 5 per cent p.a. on that capital. As far as this second requirement was concerned the amount was only small, of the order of about £3,000, for in the company's five years of operation the last three had already provided an average dividend of 5 per cent so compensation would only be required for the first two, when capital was relatively small. The first requirement, however, was going to cost the Corporation dear, for judging by normal issues, stock of $2\frac{1}{2}$ per cent would be issued, thus requiring £2 of stock for every £1 of capital. Hence each shareholder stood to take a handsome 100 per cent profit on his investment, although as they pointed out – 'we took the risk at the start'.[4] As current investment stood at about £160,000, the Corporation faced a bill of about a third of a million pounds for the purchase of the undertaking.[5]

On Wednesday 1 September the Council debated the issue at great length. Mr. Ford, the presenter of the resolutions, laid many facts and figures concerning the electricity company before the Council to prove the necessity of a takeover, pointing out that as the company was rapidly expanding, and thus increasing capital expenditure, delay would be very expensive. To mollify the fears of those who still worried about possible annual losses if the Corporation were in charge of the undertaking Mr. Ford referred to the fact that there were now 225 electric lighting works, 159 of which were owned by public bodies. He did not think that anyone could point to a failure in the Corporation management of electric lighting and exampled particularly Bradford, Dewsbury, Halifax, Huddersfield, Hull and Manchester. All these towns had had municipal control from the start, and were successful; Liverpool had taken over control from a private company and had made satisfactory surpluses.[6]

Mr. Ford was congratulated by other members for his very lucid statement, but it was considered that at that late hour it was impossible to grasp the question in all its bearings. It was therefore suggested that the whole matter be turned over to the Parliamentary Committee to inquire and report. (It was during discussion on this suggestion that an amusing incident occurred. Ald. Gordon, who incidentally was the auditor and a small shareholder of the company, was in the middle of his comments when the electric lights almost went out for a minute or two. After pausing to allow the lights to brighten and for the laughter to subside, Ald. Gordon expressed the hope that this was not an ominous sign!)[7]

In view of the various difficulties the Council soon agreed to let the Parliamentary Committee deal with the matter. And it looked as if they had a problem on their hands, considering Ald. Gordon's opinion on the view the company would take on the interpretation of the term 'annuity.'[8] Yorkshire House-to-House believed annuity to mean an annuity in perpetuity, which could only be provided by the issue of irredeemable stock. Normal corporation stock was redeemable at a specific date, when it was possible that new stock of lower income could be issued, and this, in the company's view, was not acceptable. The problem facing the Parliamentary Committee was that the Corporation was no longer able to issue irredeemable stock, having foregone this power as one of the conditions laid down by the Inspector of the Local Government Board in March 1895 when he allowed the Corporation power to borrow £100,000.[9] This meant that if Yorkshire House-to-House were correct in their interpretation, the Corporation had no legal way to buy the company! What a problem – but how typical!

The Sheffield predicament

The Parliamentary Committee were fortunate, to some extent, that they were not ploughing a lone furrow, for by chance the Corporation at Sheffield were in exactly the same predicament as that at Leeds, the only difference being that they were some six months ahead, having applied to take over the electricity company in their city in April of that year, 1897.[10] The case at Sheffield was possibly more ironic than at Leeds, for the Sheffield Council had soon discovered that the Act removing their power to issue irredeemable stock and the confirmation of the Order establishing the electricity company had been given Royal Assent on the very same day.[11]

The situation at Sheffield was further complicated by the fact that three of the Councillors were also directors of the company and several other Councillors – including the Mayor – were shareholders. The Town Clerk ruled that these Councillors were not eligible to vote on matters concerning the takeover – indeed they were often required to retire from the Council Chamber – and he gave a similar ruling about those Councillors having a financial interest in the Gas Company.[12] This effectively ruled out half the Council from meaningful decision, so there was little opposition to a recommendation (later claimed to be wasteful and expensive)[13] from the Parliamentary Committee to refer their problem to the law, and ask for a ruling from the Courts.

The hearing began on 3 December 1897[14] in the Chancery Division of the High Court, and five days later Mr. Justice North ruled that annuity did indeed mean 'in perpetuity' and required the issue of irredeemable stock, which he agreed the Corporation were now unable to do.[15]

The *Sheffield Independent* of 8 December observed that the Corporation, now deprived of any statutory powers to fix the price, were at the mercy of the company, an observation surely received with satisfaction by the directors of Yorkshire House-to-House in Leeds!

This was not a situation which either of the Councils, Sheffield or Leeds, viewed with relish, for they could both see themselves paying more for the electricity companies than they had anticipated. As a result, while desultory negotiations continued, both Councils decided to apply for Provisional Orders (under the Public Health Act of 1875) to enable them to issue irredeemable stock.[16]

It was recognised that these Provisional Orders would be difficult to obtain, for, as Mr. Wilson (Chairman of the Parliamentary Committee) said in the Leeds Council: 'the Local Government Board, and Parliament behind them, had set their face against the issue of irredeemable stock.'[17]

This Parliamentary attitude was confirmed when the Sheffield Parliamentary Committee received a letter from the Local Government Board which stated: 'The Board . . . point out that for many years past they have entertained objection to the issue of irredeemable securities by local authorities, and they have been accustomed to represent their views in this matter to Parliament. It has not been the practice in Parliament during recent years to authorise the issue of irredeemable stock.'[18] The

Corporation would be better advised to promote a Private Bill, concluded the Board.

However, the Sheffield Parliamentary Committee were not willing to accept defeat over the Provisional Order. They recommended that the Council continue their pursuance of the Order but this time in conjunction with Leeds, and suggested a joint deputation to wait on the President of the Local Government Board.[19] At the same time they recommended renewal of negotiations with the Sheffield Electric Light and Power Company on the basis of £220 of 2½ per cent redeemable stock for every £100 of capital expended up to 29 September last.[20]

As these terms had almost been agreed before the matter had been taken to law, the Council decided to accept the recommendation to re-open negotiations, a decision taken on 12 January 1898, but soon found that the company had decided on a takeover date of 31 December 1897. The Company knew full well the strength of their position and were determined to stick to as late a date as possible; the later the date, the more the capital expended, and hence the bigger the profit from the Corporation.

It didn't take Sheffield Council long to realise that they had little bargaining power and on 9 February announced that they had agreed terms with the Company. The main provisions were of course the £220 of redeemable stock per £100 of capital, and a purchase date of 31 December 1897. Other provisions included the payment by the Corporation of £5,726.12s.10d for the company's stock-in-trade and stores; a sum equal to 10 per cent (free of income tax) on the share capital from 31 December 1897 to the date fixed for completion of purchase; purchase of the book debts by the Corporation; and an agreement that the company could keep the final year's profits, a sum of £10,562.[21]

The terms of purchase were not universally acceptable to members of Sheffield Council and there were many harsh comments in Council, many publicised in the Press. Some described the terms as 'monstrous' and 'iniquitous' and many of the Council members admitted that the terms were 'onerous in the extreme'.[22] These accusations were mirrored by the *Sheffield Daily Telegraph*, which on 11 February commented: 'That agreement is a striking document. "Pay here," is writ large over the whole of it.'

These sentiments were not welcomed by the compnay, and four days later they let it be known that in view of the many rude comments and the 'certain reflections made upon the good faith of the directors of the Company' they had withdrawn the offer, in an obvious fit of pique.[23]

The directors then invited the Council to inspect the records and accounts to satisfy themselves of the reasonableness of the company's conditions. They also, with supreme arrogance and condescension, intimated that they might consider re-opening negotiations, subject to suitable apologies from the Council.

A worried Parliamentary Committee soon met and passed an ingratiating resolution expressing the opinion that 'The directors are fully acquitted of any improper secrecy or conduct,' and they further expressed the hope that the terms agreed by the Council would again be accepted by the company.[24]

There is no doubt that apart from the feelings of personal anger at the attitudes of Council colleagues, the Directors of the Electricity Company must also have wondered if they couldn't improve the purchase terms. They did, after all, hold the whip hand and had so far got what they wanted, but this was an idea which did not last for long. The *Sheffield and Rotherham Independent* was no doubt near the truth when it wrote: 'the company accepted the very fat and substantial bird in the hand and finally surrendered all envious thoughts about birds in the bush that were only problematically fatter.'[25] On 24 February 1898 final agreement of purchase was announced.[26]

It now became necessary for Sheffield Council to obtain approval from the ratepayers to petition for a Bill in Parliament to authorise purchase of the company and after public advertisements in the streets and newspapers a public meeting was held on 26 April.[27]

Unfortunately for the Corporation local electrical contractors had ensured themselves a large representation among the 100 persons present at the meeting and they objected strongly to the fact that after purchase the Corporation would be able to continue the trade of wiring consumers' premises and making and fixing consumers' fittings, as done by the company. The Lord Mayor would not give an undertaking that the Council would not carry on this trade and as a result the meeting voted against the takeover, a decision which committed the Corporation to a much larger poll of the ratepayers. Even those who were against the purchase felt that this was an unnecessary extra expense of £1,000 (which was the estimated cost of the poll),[28] and when 93 per cent of the returns showed in favour of purchase[29] there was considerable complaint about wasted money. However, the Council now knew they had the support of the ratepayers. There were no barriers remaining to completing purchase.

The Leeds takeover

The Leeds Council watched the events in Sheffield with interest, which rapidly turned to apprehension as the cost of the takeover was revealed. Like Sheffield, Leeds too had intended to continue negotiations while at the same time pressing ahead for a Provisional Order, and it soon became obvious that unless a Provisional Order were obtained, Leeds Council would also be forced to pay a heavy price for the electricity undertaking in their city.

The directors of Yorkshire House-to-House considered an expensive takeover as a bird in the hand but unlike their colleagues in Sheffield began to work for fatter birds in the bush. Their first approach was to attempt to persuade the Council that a takeover was not appropriate at this time. The plant might soon be rendered obsolete by changes in electrical science, they argued, and it would be advantageous to wait. The Council were not foolish enough to accept this argument, knowing full well that if the plant were to be renewed, the capital – and hence the takeover price – would be vastly increased. Nor were they willing to accept the company's second argument, that there would only be small return on the capital.[30] The *Leeds Mercury* spoke for most of the Councillors when it commented on 6 January 1898: 'The question of the return to be obtained for the enterprise during the first few years is of small moment. The Corporation are providing for the light of the future, and can afford to await the fuller development of this new source of revenue.'

In any case, there was every prospect of very good returns, for the company had of course done exceptionally well. The *Yorkshire Evening Post*, on 12 January pointed out that 'the great success of the company must make some people feel inclined to kick themselves.'

Having thus been assured that the Council were serious in their intentions, the next step for the company was to persuade the Council that an agreed sale, rather than a statutory sale, would be beneficial to all parties. On 1 February, the Chairman – Grosvenor Talbot – told shareholders that he did not think the company would wish to drive the Corporation unnecessarily to seek powers under a Bill at the cost and annoyance of the people of Leeds. He looked forward to negotiations with Council representatives.[31]

Three weeks later the Leeds Parliamentary Committee recommended a definite offer to the company of £210 redeemable 2½ per cent stock for every £100 capital.[32] It's puzzling how their negotiations with the company could have persuaded them that this was a realistic offer. 'Seeing that the Sheffield company succeeded quite recently in obtaining much better terms,' said an editorial in the *Yorkshire Evening Post*, 'I hardly imagine that Mr. Grosvenor Talbot and his friends will jump at these.'[33]

And so it proved. On 11 March the company rejected the offer, commenting that terms similar to those at Sheffield would be more appropriate.[34]

The directors of the company were quite sure that they held the whip hand, so it must have been a considerable surprise to learn that the Local Government Board had agreed to hold a meeting of inquiry in Leeds.[35] After all that the representatives of the Board had said in the past, this decision was extremely unexpected and it was due apparently to the efforts of the Leeds MP Mr. Jackson who, according to the *Yorkshire Post*, 'was interested in the matter, and knowing the right door to go to and being able to gain access to the right man, he had smoothed away all difficulties.'[36]

The inquiry was fixed for 31 March. It was unremarkable except for the fact that the lawyer representing Yorkshire House-to-House repeated the company's assurance that the company were 'anxious to do what appears to be reasonable'[37] and it was no surprise (at least to Leeds Councillors) to hear that the application had been successful. Half-way through May the Local Government Board informed Leeds that the Provisional Order had been made and would shortly be submitted to Parliament.[38]

This was indeed good news to the Leeds Council and the local press, too, were delighted. Particularly satisfying was the calculated saving of £40,000 compared to what the cost would have been on Sheffield's terms, a fact reported rather ruefully in Sheffield.[39] There was some apprehension in that city, an apprehension which caused one worried correspondent to the *Sheffield Daily Telegraph* to ask if this was true.

'Your readers will anxiously await further light on the subject', he concluded, 'unless they are content for their city to be looked upon as SHEFFIELD, NEAR LEEDS.'[40]

Sheffield Council, for their part, did not believe that Leeds had done all that well, in fact they were of the opinion that the Leeds terms would prove much more expensive than anyone anticipated. The problem at Leeds, they explained, was that a sinking fund would have to be established to buy back the irredeemable stock as it became available on the market, and for that purpose the Corporation at Leeds would be empowered to issue redeemable stock. But, claimed Sheffield Council, the sinking fund would not be sufficient, for the irredeemable stock would be expensive. 'If they expected that they would be able to redeem it at anything like par they were very sanguine,' reported the *Sheffield Daily Telegraph*, and in fact it was considered that the stock could easily double in value. There would obviously be a heavy burden on the rates.[41]

This burden was considered in a different way by *The Statist* a few months later (on 19 November 1898). This magazine assumed that the purchase price would be £220,000 – not far off the eventual price agreed – and argued that the assets thus purchased would have to provide enought profit to pay the 5 per cent interest on the stock and also to contribute to a sinking fund so that the interest could continue when the life of the assets had expired. This would involve a payment of £11,000 p.a. for interest and £5,600 p.a. to the fund.

The Local Government Board had sanctioned a redemption period of forty years which *The Statist* thought was 'quite astonishing'; the normal period of depreciation was in the order of twenty-five years and this period would be much more appropriate. To make this reduction it would be necessary to write off a large proportion of the capital each year, and this would take another £5,800 p.a. The total annual charge for interest, sinking fund and reserve would thus come to £22,400.

The Statist was full of doubts. 'Whether such profits can be earned on the low prices for electrical energy now charged by the company,' it pondered, 'is a problem that must be left to the engineers to solve.'

None of these arguments counted for much with the Council in Leeds, but they were worried at the hazards of the journey of their Bill through Parliament. The longer it took to have the Provisional Order confirmed the more the purchase would cost. The electricity company were of course aware of this and in June they made a last ditch effort to prevent the confirmation of the Order. They insisted on representation before the Select Committee which was considering the Provisional Order and when the Corporation objected the matter had to be referred to the Court of Referees.[42] The company here made two cases for being heard before the Committee. First they claimed that the Order made by the Local Government Board was *ultra vires* (beyond their power), as the Public Health Act 1875 specifically provided that no powers should be conferred on persons for their individual benefit, and yet the stock to be issued by the Corporation would be redeemable by individual agreement. They also claimed that it was no longer competent for the Local Government Board to bestow powers which had been abrogated. Such a thing could only be done by Parliamentary Bill, after consultation with the ratepayers (as at Sheffield). But the Court denied the company a *locus standi,* probably to the intense relief of the Corporation, and the Provisional Order Bill once more continued its progress through Parliament.

Nerves were now becoming strained and only a few weeks later considerable disquiet was expressed in Leeds Council about the activities of the company. There was a suspicion that the company was unduly increasing its capital expenditure by unnecessarily laying large lengths of new mains. At the Council meeting early in July one Councillor drew attention to what he described the 'unseemly haste' of the company in laying mains, which he claimed were being 'chucked down' in a negligent way.[43] The *Leeds Express* agreed that the apparent hurry was 'not very seemly' and hoped that the directors (who were 'men of local eminence and repute') would consider the remarks made in Council.[44]

The remarks were indeed considered, and very rapidly, for the very next day there appeared in all the Leeds newspapers a letter from Grosvenor Talbot, the Chairman of the electricity company. In this he strongly refuted all allegations. He asserted that the work was being done to a high standard, and in fact was being supervised by representatives of the manufacturers, and explained that it was quite normal to do this sort of work in the summer, when connections could be made with minimum disturbance to existing consumers. The only difference this year was in the scale of the work, for the continued success of the company had resulted in a massive increase in lights of almost 40 per cent and hence a much increased requirement for new mains compared to previous years.

These assurances quietened the criticisms, and fortunately there was little time for fresh complaints to arise, for at the end of July it was announced that the Bill confirming the Leeds Provisional Order had at last received Royal Assent.[45] The Town Clerk made this announcement to the Leeds Parliamentary Committee on 28 July and there was an immediate decision to ask the Council at its next meeting to give notice in writing of its intention of purchasing the electricity undertaking.[46] The Council needed no persuasion in this matter, of course, and having formally expressed its intention of taking over the electricity undertaking[47] it asked the Parliamentary Committee to form an Electricity Sub-Committee to carry out the negotiations.

Those who believed the affair was now all but concluded must have been surprised at the resistance still shown by the company, for the first meeting of both parties was lengthy – more than three hours – but with little result. Even though the terms of purchase of the capital assets was laid down in the Provisional Order, the definition and extent of the assets, and when purchase would take place, were but a few of the many loose ends still to be tidied up.[48]

This first meeting resulted in an agreement that the date of takeover of the capital assets would be 1 September, from which date the company would work the undertaking purely on behalf of the Corporation. Takeover of the revenue expenditure would be 30 September, the end of a quarter and a convenient time to balance the books. It was hoped that the purchase would be completed by 15 November. For the meantime it was agreed[49] that two Councillors would join the directors on the Board of the company – to protect the Corporation's future interests – while accountants from both sides would inspect the accounts to determine the actual capital.

Over the next few months agreement was reached on the outstanding details. The most difficult clause to arrange was the interpretation of what actually constituted 'capital assets', and these were eventually defined as 'all goods and machinery actually delivered or goods manufactured under contract or order of the company, and finished in the works of the contractors on 31 August.'[50]

It was also established that the Corporation would take over the whole staff of the company, their agreements for service, all contracts, and fire and other insurance policies. The Directors agreed to remain in post from 1 October to the final purchase date, the fees to be provided by the Corporation.[51]

On 8 November the full agreement was signed by both parties: the Chairman of the Parliamentary Committee and the Town Clerk for the Corporation, and the Company's Chairman and Solicitor.[52] The very next day the full Leeds Council approved the agreement and in an obvious mood of relief and conciliation there was also an acknowledgment made of 'the courtesy of the directors' and the fair and reasonable views they had taken with reference to many debatable points. Previous disagreements had been forgotten![53]

A fortnight later, at an Extraordinary General Meeting of the Yorkshire House-to-House Company on 20 November, the shareholders also sanctioned the agreement, although it must be admitted that there were not the same expressions of satisfaction that had been evident among the Councillors. Indeed the Chairman commented ruefully that the granting of the Provisional Order 'was a departure from the position the Board of Trade took in their decision of the Sheffield case a short time before – it was a grave reflection on a great Government department that they should reverse the decision made a few months previously.'[54]

By this time the joint accountants had agreed that the capital expended up to 31 August 1898 had been £217,420.10s.4d, to which would have to be added various other items to complete the purchase. These items consisted of £773.15s.2d for stock in trade and stores, £529.19s.0d for unpaid electricity bills (which would be paid to the Corporation in due course), £1,088.11s.0d for capital expended from 31 August, £94.10s.5d for directors' fees since 1 August, and finally £2,666.0s.3d for interest to the company on the money which should have been paid on 1 August. These items added to a total of £5,153, and would be paid to the company in cash.[55]

It seemed now that nothing prevented immediate completion of the takeover, but the company still seemed reluctant, so that at the Council Meeting on 7 December it was agreed that the Corporation be authorised to issue irredeemable stock at once to the required amount. This was so that the Parliamentary Committee would be ready the next day to conclude the business; the company would not be able to find any valid or reasonable excuse for putting the day off any longer. As the Chairman of the Parliamentary Committee pointed out, the Corporation had never been on a level with the company in the negotiations and 'the present seemed an opportunity to get on level terms.'[56]

The Company accepted, at last, that terms were level; on 15 December 1898 completion of the purchase took place at the company's premises in Whitehall Road. The company's works and premises were formally handed over to Councillor J. Green Hirst, Deputy Chairman of the Lighting Committee, on behalf of the Corporation.[57]

Notes and references

1. *Leeds Mercury,* 28 August 1897.
2. *Leeds Express,* 28 August 1897.
3. *Yorkshire Post*, 28 August 1897.
4. *Leeds Daily News,* 30 August 1897.
5. *Yorkshire Post,* 2 September 1897.
6. *Ibid,* 2 September 1897.
7. *Leeds Daily News*, 2 September 1897.
8. *Yorkshire Post,* 2 September 1897.
9. *Leeds Mercury,* 20 March 1895.
10. *Ibid,* 6 July 1897.
11. *The Times,* 8 December 1897.
12. *Sheffield Daily Telegraph*, 13 January 1898.
13. Editorial comment, *Sheffield Daily Telegraph,* 13 January 1898 – 'the money spent on the law-suit against the company might just as well have been thrown into the street.'
14. *Sheffield Daily Telegraph*, 4 December 1897.
15. *The Times,* 8 December 1897.
16. *Sheffield Independent,* 23 December 1897 and *Yorkshire Post,* 6 January 1898.
17. *Yorkshire Post,* 6 January 1898.
18. *Sheffield Telegraph*, 10 January 1898.
19. *Yorkshire Post,* 13 January 1898.
20. *Sheffield Daily Telegraph,* 13 January 1898.
21. *Sheffield & Rotherham Independent,* 10 February 1898.
22. Editorial, *Sheffield Daily Telegraph,* 10 February 1898.
23. *Sheffield & Rotherham Independent*, 16 February 1898.
24. *Ibid,* 16 February 1898.
25. *Ibid,* 22 March 1898.
26. *Sheffield Daily Telegraph,* 24 February 1898 – This was agreement by Sheffield Council; the Company shareholders agreed to sell at a special Meeting held on 21 March.
27. *Sheffield & Rotherham Independent,* 27 April 1898.
28. *Ibid,* editorial.
29. *Ibid.* 25 May 1898.
30. *Leeds Mercury,* 6 January 1898 – arguments put forward by Ald. Lupton and Ald. Gordon at a meeting of the Council. They were also a shareholder and the auditor of the Yorkshire House-to-House Co.
31. *Yorkshire Post* and *Leeds Mercury*, 2 February 1898.
32. *Yorkshire Post,* 3 March 1898.

33. *Yorkshire Evening Post,* 26 February 1898.
34. *Yorkshire Post,* 12 March 1898.
35. *Ibid,* 21 March 1898.
36. *Ibid,* 27 May 1898.
37. *Ibid*, 1 April 1898.
38. *Leeds Mercury,* 19 May 1898.
39. *Sheffield Daily Telegraph,* 28 May 1898.
40. *Ibid,* 3 June 1898.
41. *Ibid,* 9 June 1898 – report on Sheffield City Council meeting, and editorial.
42. *Leeds Mercury,* 22 June 1898.
43. *Yorkshire Post,* 6 July 1898.
44. *Leeds Express,* 7 July 1898.
45. *Yorkshire Post,* 27 July 1898.
46. *Leeds Mercury,* 29 July 1898.
47. *Ibid,* 4 August 1898.
48. *Yorkshire Post,* 26 August 1898.
49. *Ibid,* 8 September 1898.
50. *Leeds Evening Express,* 21 November 1898.
51. *Yorkshire Post,* 10 November 1898.
52. *Ibid,* 9 November 1898.
53. *Leeds Mercury,* 10 November 1898.
54. *Yorkshire Post,* 21 November 1898.
55. *Ibid*, 2 December 1898.
56. *Ibid,* 8 December 1898.
57. *Leeds Mercury,* 16 December 1898.

Chapter 18

Liquidation

Now that the electricity undertaking was in the hands of the Corporation, there was little justification for the Yorkshire House-to-House Electric Company remaining in business; it was almost time for the Directors to take their money and run.

First, though, there were one or two formalities to be observed, the principal one being the meeting of shareholders which was held on 24 January 1899.[1] Most of the evening was devoted to what would turn out to be the final Annual General Meeting of the Company, although this was perforce much shorter than previous such meetings. This was partly because the company's year had finished on 30 September and the busiest fourth quarter of the year was missing, so there was little to discuss. Brevity was also due to the fact that there was no future to consider.

Despite the briefness of the Annual Report it revealed heartening figures which showed that the company had prospered to the end. Lamps connected to the system had increased in the nine months by a third, from 49, 150 to 65, 328 and the profit was £10, 740.5s.10d,[2] most of which was dedicated to the payment of dividends which brought the equivalent dividend for the year to 6 3/4 per cent.

Towards the end of the meeting there were unanimously supported resolutions of gratitude to the staff of the company. Particular tribute was paid to the Engineer, Mr. Dickinson, and to the Secretary, Mr. Green, both of whom were awarded an honorarium of 100 guineas.

Immediately the Annual Meeting was finished, there was held an Extraordinary General Meeting, the sole business of which was to consider a resolution that the company be wound up. There was no opposition, of course, and the approved resolution was then submitted as a special resolution to a second Extraordinary General Meeting which the directors had arranged for 10 February.

This second meeting began by voting £210 out of company funds to Robert Hammond,[3] the directors expressing their wish to recognise 'the conspicuous care and skill exercised by Mr. Robert Hammond, consulting engineer, in designing and superintending the construction of the works, as well as his valuable efforts in establishing the company and obtaining the Provisional Order for supply of electricity in Leeds.'[4]

The previous resolution to liquidate the company was then confirmed and the Chairman was appointed Liquidator to be assisted by the rest of the directors. For this duty they were to be paid a year's director's fees.

The founders benefit

The main duty of the Liquidator was to sell off the Corporation's irredeemable stock, which could not equitably be divided among the sharehoulders, and by June[5] this had been done at an average price of £170.9s.11-3/4d.[6] After allotment the sale price was payable in instalments, due 3 July, 3 August and 3 September and the money was shared out as it came in.

On 21 March 1900 there was a meeting of shareholders to hear a statement from the liquidators who reported that liabilities of the company of £39, 783.10s.10d had been paid off, and all the remaining money had been distributed among the shareholders except for £18,002.3s.11d. A patentee claim was still outstanding which would take almost £3000: when this had been cleared there was expected to be £15,219.18s.2d remaining for distribution.

The gross proceeds of the sale – £370,591 – were therefore reduced to a net pool of £328,025, which because of the priority of the founders shares was allocated in an amazing fashion, and much to the benefit of the directors. The pool was first dedicated to repaying the amount paid up on the shares, a sum amounting to £169,240, there being 100 founders and 33,748 ordinary shares, all £5 paid up. The remainder of the pool – £158,785 – was then equally divided, half to pay a bonus on the 100 founders shares and half as bonus on the ordinary shares. Thus each founders share became entitled to a bonus of £794, or nearly 16,000 per cent, whereas the ordinary shares had an entitlement of only £2.7s.0d per share, a bonus of only 47 per cent. This division put the directors in a very superior financial position, of course, for the seven of them had retained half the founders shares, we may recall, and held about a fifth of the ordinary shares between them. The allocation of the proceeds of liquidation meant that not only did they have returned to them their initial investment (of about £34,000) but they also received a bonus of nearly £56,000; within the short lifetime of the company they had made a profit of almost 160 per cent, and that was ignoring the 5 per cent which they had received each year since 1892 (made up where necessary by the Corporation under the terms of the transfer). It's not hard to understand the opinion expressed in *The Statist* that the size of the bonuses were such as to make one 'envious of the position of the founders of the company.'[7]

In view of the benefits accruing to the directors, it's not surprising that there had been allegations of frantic and unnecessary capital expenditure during the company's final summer. There must have been many who realised that for every £1000 of capital expenditure in the last few months the holders of the founders shares would receive (eventually) £8.10s per share.[8] But accusations of fraud, or sharp practice perhaps, were surely completely unjustified, for no sooner had the

Corporation taken over the company than an investment of £44,000 was announced (for the immediate provision of new boilers, engines, substations and mains),[9] with a probable extra expenditure of £38,000 within the next couple of years.[10] The Lighting Committee – who had now taken over responsibility for electricity supply – argued persuasively (just as the company directors had) that 'the growth and demand for electricity was very rapid, and the Engineer advised that the work should be pushed on.'[11]

Output doubles under Council control

A couple of months later it was announced that £80,000 had either been spent already or was shortly to be so, and another £12,000 had been put aside[12]. The result was a near doubling in the output, and the existing premises being now full, there was a request for Council approval for the expenditure of another £58,000 to develop the adjacent Brittania Mills site, owned now, of course, by the Corporation. Of this sum £25,000 was required for buildings and £33,000 for machinery and mains. Demolition of the site premises began in August 1899.[13]

The major problem now facing the Council was what sort of electrical system to decide upon for the new equipment, for there had been an increasing realisation that the existing single phase system was not ideally suited to give supply for power purposes, to drive electric motors.[14] The main difficulty, as explained by the Works Engineer (Mr. Dickinson) was that 'a single-phase motor cannot be started against a load. Like a gas engine it must be commenced light, which in many instances is decidedly inconvenient'.[15] As the major extensions of mains planned by the Corporation were towards the industrial area of Hunslet[16] an inability to provide electricity suitable for power supply purposes was seen as a serious deficiency, so the Lighting Committee undertook two courses of action: they despatched the Works Engineer and the Chairman of the Committee to America to investigate the latest technology;[17] and they commissioned a report from two famous electrical engineers, Messrs. Hopkinson and Talbot.

The report discussed the relative merits of using DC – restricted in transmission but ideal for motors – three-phase AC and two-phase AC and recommended that the three-phase system would be most suitable.[18] These alternatives provoked some argument which was only resolved when the Lighting Committee decided that use of the three-phase system would commit the Corporation unnecessarily to the American firm of Westinghouse.[19] The Works Engineer admitted that three-phase was probably a superior system but was confident that the two-phase system would be adequate. As it was explained to the Press later, 'such a system is in operation . . . in numerous parts of America. In the Niagara Falls Scheme the electricity is generated at two-phase and distributed at three-phase. The conversion of the two-phase to three-phase does not appear to be a very expensive matter, supposing, which Mr. Dickinson thinks most improbable, the scheme of the Corporation does not answer all requirements.'[20]

The Council accepted the advice of the Lighting Committee at its meeting held on 8 February 1900[21] and later on there was an agreement that not only the new plant on the Brittania Mills site, but also the plant in the old works should run at two-phase. In May 1900 tenders were invited for 'two or three' 630 KW combined engine/alternator sets[22] for replacing sets in the old works. It was explained that the plant in question was 'not absolutely worn out, but needed replacing on account of the advance of modern science. So rapid, indeed, has this advance been during recent years, that the old electrical machinery, though the best known when put down, is now no longer regarded as efficient.'[23] The Council subsequently agreed to the purchase of three sets.

Having thus decided on ways of providing electricity for motive power, it then became necessary for the Council to give urgent consideration of the effects of this provision on their plant and their finances. By late 1900 there had already been a 'striking growth'[24] of the undertaking since its takeover by the Corporation – plant capacity had increased from 2400 to 4300 kW (since the commissioning of the previously ordered plant) and there had been an increase of more than 80 per cent in the number of lights connected to the system (now 118,111). With the availability of a two-phase system there was a growing recognition by local trade and industry of the 'cleanliness and convenience of electric motors'[25] which seemed likely to impose new difficulties on the Corporation's attempts to keep pace with the demands of the consumers. There would certainly, for instance, be a demand from industrialists for rapid extensions of the new two-phase system to their areas; there would eventually be a need to convert the whole system to two-phase and this would probably involve expensive alterations to the new sub-stations. As the new system proved its worth and grew in popularity, as it surely would, there would then be an even greater pressure on the generating capacity of the station on Whitehall Road. There was every prospect, therefore, of a much greater financial commitment by the Corporation than had been envisaged when the undertaking had been taken over, and this at a time when the Corporation had already spent £141,930 on new works, borrowed on bank overdraft.[26] It now became necessary to formalise the Corporation's financial arrangements and after a meeting in Leeds with the Local Government Board Inspector[27] they were given sanction to borrow £500,000 (later increased to £722,820)[28] 'for the purposes of electric lighting.'

The Lighting Committee were now able to continue their work without hindrance and in August 1901 there was a report[29] that the three new 630 kW machines were almost ready for running in the old station, three 100 kW sets having already been removed. A year later the new station on the site of Brittania Mills had been completed and on 30 September the members of the Council were invited by the Chairman of the Lighting Committee to inspect the plant and works just before the final commissioning by the contractors.[30] The Councillors must surely have been impressed by the two huge steam engines, both rated at 2400 h.p., to which were directly coupled the alternators rated at 1500 kW which, incidentally, supplied current at a frequency of 50 c/s rather than the 83 c/s which the Corporation had inherited from Yorkshire House-to-House. One of the engines had been supplied by the Leeds firm of McLarens, the other by Bellis and Morcom of Birmingham, and although both engines were very similar the Bellis engine was proudly described as 'the most potent engine of its kind that has yet been constructed.'[31] The Council members were inordinately proud of those engines, and of the other new equipment, which they considered would 'rank among the finest in the Kingdom'[32] – there were those, indeed, who claimed that the plant would even be 'the envy of the Americans'.[33] The pride of the Councillors was somewhat understandable for this was the first public supply station for which they had been wholly responsible and after its commissioning in the autumn of 1902 it added a huge 3000 kW capacity to the Corporation's generating capability. The increase was not extravagant, however, despite the fact that the capacity of the works was now, at 8470 kW, more than treble the installation which the Corporation had taken over.

Fig. 35. *Frontage of 'City of Leeds Electric Lighting Works 1902', in Whitehall Road. Substantially unchanged in 1984 (Yorkshire Electricity Board archives).*

There were fears expressed, prior to the commissioning, that the plant would not be sufficient for the immediate purposes, that the energy from the new buildings would 'probably be mortgaged'.[34] In the event, though, the generators were adequate for the maximum load installed in that winter of 1902–3, load which was stated to be the equivalent of 231,734 35-watt lamps, or 8110 kW.[35] This too – like the works capacity – represented an increase of more than 300 per cent during the four years of municipal control, and the units sold that year (4,448,650) were almost four times the units sold in 1898.

There is no doubt that this spectacular growth over the four years was due almost completely to the expansion of the lighting load and indeed the Council were happy to call the new station their 'Electric Lighting Works'[36] but this was a situation which was changing rapidly. Whereas in 1898 there were six consumers taking electricity for power purposes, with a tiny aggregate installation of only 20 h.p.,[37] at the end of 1902 the number of power consumers had increased to 243, their motors totalled 1,436 h.p. and 1 million units (or 20 per cent of the total units sold) were for power supply.[38] At the beginning of 1910 the number of customers using motors had grown dramatically to 1,127, with an aggregate installation of nearly 13,000 h.p., and a consumption during the year of almost 7 million units, which represented more than 55 per cent of the total current supplied.[39]

Fig. 36. *Two steam engines and alternators in new Leeds works 1902 – note room allowed for future expansion (Yorkshire Electricity Board archive).*

It was about this period, too, that the industry nationwide experienced major growth in the field of lighting, due to the introduction of the new metal filament electric light bulbs. These were much brighter than the old carbon filament bulbs and were more efficient, too, so that the customer received a better light at less cost. Understandably, the new bulbs were instantly popular, so that during the year 1909 the number of customers for the electric light increased in Leeds by nearly 394 (7 per cent) which was 'greater than for several years past'.[40] The following year the number of lighting consumers increased by 629. This increase was described as the greatest on record for the last six years.'[41]

Triumph over adversities

It's quite clear, then, that the commissioning of the new station in 1902 marked a clear 'watershed' in the history of electricity supply in Leeds. The successful operation of the new plant and its connected two-phase 50 c/s systems of cables

Fig. 37. *Bellis-Morcom engine, Leeds works, 1902 (Yorkshire Electricity Board archive).*

and sub-stations seemed to confirm that there were no longer any barriers remaining to prevent the wholesale extensions of the network throughout the city; at the same time there was a growing demand from commerce and industry to avail themselves of the new and effective source of energy for motive power, and the electric light was becoming more and more popular as the public accepted its new brightness and economy. In other words, the opening of the new works represented the start of a confident, sucessful and popular enterprise, in contrast to the doubtful and speculative industry which had existed prior to the turn of the century (technically if not financially in Leeds, perhaps). It must have been clear to the Leeds Council – particularly to the Lighting Committee – as they commissioned their new station with such pride in the autumn of 1902 – that the electricity industry was at last here to stay. The electric light had finally triumphed over the adversities of politics, technology and finance. They were right of course, although it's doubtful if they realised how right: surely even the most far-sighted Council member would have been amazed to discover just how important electricity would become in our lives more than three-quarters of a century later. How surprised they would have been to see how electricity would reshape industry and entertainment, would revolutionise commerce and medicine, and would make possible the miracles of electronics and the computer. And yet those Leeds Councillors of so long ago would have shared with us the recognition that our modern electrical world was only possible because of all the experiment, development and investment in electric lighting systems which took place during the nineteenth century. We all have cause to be thankful to those who worked so hard for the success of the electric light.

Notes and references

1. *Leeds Mercury*, 25 January 1899.
2. *Yorkshire Post*, 16 January 1899.
3. *Ibid*, 11 February 1899.
4. *Electrician*, 20 January 1899.
5. *Electrician*, 16 June 1899.
6. *Leeds Mercury*, 22 March 1900 – 'An arrangement was made with Messrs. F. Banbury & Sons, London, whereby the firm offered the stock for sale by tender at the minimum price of £170 per cent, which minimum price was guaranteed by them. The issue was over applied for.'
7. The *Statist*, 19 November 1898.
8. Each £1,000 expended would entitle the company to receive Corporation stock worth, eventually, £1,700. This would be divided in the proportion £850 for the 100 founders shares, £850 for all the ordinary shares.
9. *Leeds Mercury*, 31 December 1898.
10. *Yorkshire Post*, 5 January 1899.
11. *Ibid*, 5 January 1899.
12. *Leeds Mercury*, 13 March 1899.
13. *Yorkshire Post*, 30 August 1899.
14. *Leeds Mercury*, 13 March 1899.
15. Mr. Dickinson, Works Engineer – *Yorkshire Evening Post* 4 March 1901.
16. *Yorkshire Post*, 30 August 1899.
17. *Ibid*, 13 March 1899 – Council Meeting: 'Ald. Wigram having to leave early for America...' and 30 August 1899 'The Engineer of the works is just returning from the United States, where he has been making inquiries as to the different systems of generation and distribution of electric light.'

18. *Leeds Mercury,* 3 February 1900.
19. *Yorkshire Post,* 3 February 1900.
20. *Ibid,* 9 February 1900.
21. *Ibid,* 9 February 1900.
22. *Electrician,* 11 May 1900.
23. *Yorkshire Post,* 16 May 1900.
24. *Ibid,* 20 October 1900.
25. *Yorkshire Evening Post,* 4 March 1901.
26. *Leeds Mercury,* 20 April 1901.
27. *Ibid,* 20 April 1901.
28. *Yorkshire Post,* 28 September 1904 – sanction to borrow £505,400 was obtained in July 1901 and a further sum of £220,000 was sanctioned in December 1903.
29. *Leeds Mercury,* 20 August 1901.
30. *Ibid,* 1 October 1902 – 'The new engines have not yet been used for the supply of the city, as the contract time has not expired.' Strangely, in view of the Council's pride in this, their first new station, there appears to have been no official opening ceremony and there seem to be no press reports of the contractors handing over the station to the Corporation.
31. *Birmingham Daily Post,* 28 May 1902.
32. *Leeds Daily News,* 30 September 1902.
33. *Ibid,* 30 September 1902.
34. *Yorkshire Evening Post,* 3 February 1902.
35. *Yorkshire Post,* 17 June 1903.
36. The facade of the 1902 station still stands at the Whitehall Road site, not far from City Square. Picked out elegantly in the brickwork above the two double-gated entrances are the words: CITY OF LEEDS ELECTRIC LIGHTING WORKS 1902.
37. *Yorkshire Post,* 28 September 1904.
38. *Ibid,* 17 June 1903.
39. *Ibid,* 21 June 1910.
40. *Yorkshire Observer,* 21 June 1910.
41. *Ibid,* 17 May 1911.

Epilogue

The demise of the Yorkshire House-to-House Electric Co. could never really have been considered to be a sad occasion, for it is hard to think of anyone who suffered from the closure. The shareholders, for example, after receiving good dividends during the life of the company, then got all their money back; the directors all made small fortunes at the liquidation; the Corporation took over a thriving and expanding undertaking; and the consumers looked like getting an even better service than before. Even the employees had no complaints, for all of them kept their jobs, under the terms of the takeover, and many were given an increase in wages to put them on a level with comparable staff at the other Corporation station at Crown Point.[1]

The only group who received the news of the company's liquidation with a degree of regret was the parent House-to-House company in London which, as we saw earlier, had been formed on the initiative of Robert Hammond in the hope of establishing a nationwide group of House-to-House stations.

The fortunes of the parent company

Despite a healthy and increasing volume of business by this parent station – which by 1898 had attracted about 1,500 consumers and was making annual profits of more than £12,000[2] – the pressure on the Leeds company to sell out to the Corporation came as growing evidence that the dream of expansion had ended. This only confirmed similar evidence from another source: the directors reported sadly, each year, that the business of the Leeds and London Electrical Engineering Co. (formed to build House-to-House stations) was not sufficient to enable the distribution of profits. Unhappily, in 1897, this construction company went into voluntary liquidation.[3]

The directors of House-to-House soon accepted the inevitable: there would be no spread of their system round the country. At an Extraordinary General Meeting held in August 1899 the shareholders agreed to change the name of the company from the general apellation 'House-to-House' to the more local 'Brompton and Kensington', as this was the area of the company's supply.[4]

Not that this restriction on the company's horizons detracted from the determination of the directors to run a successful business, for now, as before, the company showed the same sense of initiative and economy that had always been apparent in Leeds. As in that city, it was soon realised that as the numbers of consumers increased it was more economical to give supply at low voltage from sub-stations, rather than give each customer a separate transformer, so in 1897 the first installation of sub-stations was reported.[5]

There was always a determination by the directors to keep up with technology and there was, therefore, a constant renewal of generating plant, although this resulted in higher depreciation charges. By 1900 the rope-driven plant had been completely replaced by high-speed directly driven generators and the plant was now so standardised and reliable that the directors felt able to do away with the post of Managing Director.[6]

There was a 'hiccup' in the achievements of the company about 1909 when the effect of the new metal filament lamps was first felt.[7] These lamps were much more efficient than the old carbon filament lamps, giving more brightness with less current, an effect which contributed to a decrease in company profits from a high at the end of 1907 of £32,000[8] to £27,000 for 1909.[9] It took another two or three years for a recovery, helped by the installation of a turbine for the first time in 1911.[10]

At the same time the company took a step which was unique among electricity undertakers. In 1912 a subsidiary company was established, the Brompton and Kensington Accessories Co. Ltd., with shares wholly owned by the parent company. The purpose of this company was to publicise the benefits of electrical apparatus, particularly for cooking, and thereby to promote the use of electricity.[11] To do this the accessories firm established a factory for the design and manufacture of its own cooking apparatus and furthermore (this was what made the project unique) it established an adjoining restaurant, which soon became a popular feature of the area. It was claimed that this was the only supply company in the world which not only supplied electricity but made its own cooking apparatus and used it in the regular course of business.[12]

The First World War

Just as the Company appeared to be flourishing, however, business was interrupted by the First World War, which had various effects on both the company and the electricity industry as a whole. For the Brompton and Kensington undertaking, the war was a difficult time. Many undertakings benefitted from the fact that the shortage of manpower for industry produced a sudden demand for electricity for motive power, but this was of no benefit to Brompton and Kensington with its single-phase 83 c/s system. Instead, the company suffered because of the restrictions which the Government were forced to put on the use of electricity in order to concentrate available resources on vital productions. It was inevitable that these restrictions reduced the use of non-essential lighting and this was accentuated by the Summer Time Act of 1916.[13]

The shortage of labour had a drastic effect on wage levels too, and the company reported at the end of 1917 that wages were almost double pre-war levels.[14] Labour costs were also increased by the voluntary gratuities which the company made to dependants of employees on active service. All in all, it is hardly surprising that the directors had to reduce dividends in April 1917, the first reduction for thirteen years.[15]

As soon as the war was over Brompton and Kensington had to contend with the Electricity (Restriction of New Supply) Order 1918, which prohibited connection of supply to buildings not previously connected.[16] It was not until a couple of years had passed and these restrictions were lifted, that the company began to thrive once more. The fact that the once wealthy area of Brompton and Kensington had deteriorated slightly during the war contributed greatly to this revival, for many of the large houses were being turned into flats and maisonettes, thus greatly increasing the lighting load.[17]

Industry too parochial

For the supply industry as a whole the war only served to confirm what had been increasingly apparent since the turn of the century: the industry was too parochial, too old fashioned and too inefficient to cope with the demands of the modern industrial society. There had always been arguments about the relative merits of DC and AC systems, but the disputes had been comparatively unimportant when lighting was the only load, and when technology restricted the industry to small stations supplying small local areas on lightly loaded cables. But the increasing use of electric motors, particularly during the war, changed the whole basis of the arguments, for the motors required a polyphase AC supply.

At the same time there was a growing realisation that AC systems were ideal for long distance transmission of large amounts of power, and there were now available large generators which could provide the power economically. It was obvious that the future lay with large generating stations (probably near the coal fields) supplying consumers at long distance on high voltage transmission lines which would also interconnect the stations.

Unfortunately the reality in Britain was somewhat different. Only 14 per cent of the systems were exclusively AC[18] and interconnection was further complicated by the many different types of systems in use, both AC and DC, a confusion which the passing years did little to clear. Even as late as 1928 there were still thirty-three different AC systems and twenty-four different DC systems.[19]

There was little sign, too, of the establishment of large power stations, indeed the small stations proliferated, from 236 in 1900[20] to more than 600 at the end of the war.[21] In 1913 the average size of all the London stations was a tiny 4,670 kW and there were twenty of them with a maximum demand of less than 5,000 kW.

Considering the fact that sets of 20,000 kW were now not only technically feasible but economically attractive, it's easy to sympathise with the *Electrician*, which considered the situation to be 'ridiculous'.[22]

It is sad that the reality was so far removed from the ideal, for the effect on Britain was serious. Small-scale production resulted in expensive electricity, and during the war it resulted in insufficient electricity. The large variety of systems not only prevented interconnection but also ensured that manufacturers were not able to produce standardised equipment. This in turn led to expensive plant for the undertakers and an uncompetitiveness in world markets by British manufacturers of electrical plant.[23]

The tragedy of all this was that although the technical solutions to the difficulties were readily available, the political problems proved almost unsurmountable. Three-quarters of the undertakings were in the hands of municipal authorities[24] and in most cases civic pride prevented any thoughts of amalgamation or interconnection. Fear of private monopolies prevented absolutely any co-operation with companies, while the companies drew away from the municipal authorities because of the fear of their growing power. Despite a continuing flurry of Parliamentary activity by both sides in an attempt to rationalise and regulate the industry, mutual distrust prevented any meaningful legislation.

Attempts to improve supplies

The only real advances were the establishment by Private Act of various power companies, the purpose of which was to establish large stations so that supply could be given over a large area either to consumers not already supplied by an undertaking or in bulk to other undertakings, who would then be relieved of the burden of extending or renewing their own plant. These were not always successful ventures due to the opposition of the Municipal Authorities in their areas of supply, but one that did succeed was Yorkshire Electric Power, whose first power station was opened in December 1903 at Thornhill.[25]

In London there were several attempts to improve supplies. In 1914 the London County Council promoted a Bill which was countered immediately by a rival Bill promoted by a number of supply companies, including Brompton and Kensington. Both Bills proposed amalgamations and closures of various stations but neither were successful. The LCC Bill failed to receive adequate Parliamentary support and the companies' Bill attracted stern opposition from the Government, which considered the Bill to be of a contentious nature.[26]

Six years later the LCC revived its plans and this time Brompton and Kensington (with a total capacity of only 3,000 kW) was included in the closure plans which totalled 76,400 kW of plant.[27] Once again the plans were rejected and Brompton and Kensington was allowed to prosper, continuing to make annual profits up to about £50,000.[28]

There was no further contemplation of reorganisation, as far as Brompton and

Kensington were concerned, until the end of 1925, when a Joint Electricity Authority was established for London under the Electricity (Supply) Act of 1919. The government had intended that these authorities would have comprehensive and compulsory powers of reorganisation in the districts in which they were set up, but intense opposition had so emasculated the Act that the authorities were reduced almost to powerless advisory bodies.[29] In this case, however, the threatened influence of the London Joint Electricity Authority was immediately effective, for in March 1926, Brompton and Kensington reported that it had joined several other private companies in a consolidated system of generation controlled by the newly formed London Power Company.[30] This was not a commercial undertaking and could not make a profit;[31] it merely co-ordinated the efforts of the nine constituent companies[32] who still maintained an independent – though cooperative – existence.

The co-operation was not immediately of benefit to Brompton and Kensington, for the new interconnected generation and transmission system ran at 50 c/s and would not connect with their 83 c/s system.[33] A couple of years later, though, Brompton and Kensington's system was changed and on 15 October 1928 they started taking a bulk supply from London Power Co. at 22,000 volts, and closed down their generating station.[34] The benefits were immediate, for by the end of the year they reported a 33 per cent increase of profits for only a 14 per cent increase in sales. The Chairman commented that the savings to the constituent companies (and hence to the public) had been £450,000 that last year.[35]

In 1930 there were four power stations generating for London Power[36] and proposals had been published for the construction of Battersea power station;[37] the connection of Brompton and Kensington to this efficient system enabled the company to reduce prices drastically. In the first five years of amalgamation they were able to cut the cost of electricity by 40 per cent and there was a consequent doubling of output.[38] This growth continued over the next few years, with an average increase of about 15 per cent.[39]

The success of the London Power Co. can be gauged by the fact that in 1935 it was claimed (in the *Electrician* of 15 February) that the company was producing 10 per cent of UK electricity. This was a startling proportion, for at that time there were about 600 electricity undertakings.[40]

The demise of Brompton and Kensington

From the success of co-operation it was but a short step to amalgamation and towards the end of 1934 this was agreed by Brompton and Kensington with five other companies: Charing Cross; St. James' and Pall Mall; Westminster; Kensington and Knightsbridge; and Chelsea. The new company was called the London Associated Electricity Undertakings Ltd. and was incorporated on 2 February 1935 with a capital of nearly £7 m.[41] Amalgamation was achieved by exchanges of shares of the existing companies for stock carrying similar rights in the new company. The Brompton and Kensington Electricity Supply Co. was quoted for the last time on 30 August 1935. The following month the shares quoted were those of London Associated Electricity Undertakings.[42]

Thus ended the independent life of the Brompton and Kensington Electricity Supply Company – nee House-to-House – after a successful life of forty-seven years. Strangely, this demise was accompanied by a sad coincidence. The Chairman of the Brompton and Kensington Company since its inception in 1889 had been Henry Ramie Beeton. On 20 June 1934, just a few months before amalgamation was agreed, he died, aged eighty-one.[43]

Inspiration of Robert Hammond

Although Mr. Beeton had always directed the business of the company, the inspiration behind House-to-House had undoubtedly been Robert Hammond, who has featured regularly in our story. His interest in electrical matters extended well beyond this company, of course, and he became well known to the public. The *Yorkshire Post* thought that he was 'probably the best known electric consultant in the country . . . certainly one of the foremost men in connection with municipal electricity works.' This was certainly true in Leeds, for Hammond had been responsible for the works of Morley and Wakefield as well as those in Leeds.[44]

Hammond was, according to the *Yorkshire Post*, a man of winsome personality, with a great force of character and abundant energy, and he was also a fluent and fascinating speaker. In 1892 Hammond decided to trade on these virtues and on his popularity when he attempted to enter Parliament by contesting the Hallam division of Sheffield. He failed, in what the *Yorkshire Post* referred to as a 'minor diversion' of his career.

Although no doubt a great disappointment at the time, this failure enabled Hammond to continue devoting his time and energy to his many electrical interests, one of which was electric transport. Between 1883 and 1903 he paid several visits to the United States to study developments in this field and he acted as electrical engineer to many municipal electric tramways in the United Kingdom and the Colonies. Twice – at Paisley in 1910 and at Gravesend in 1911 – he acted as an arbitrator to fix the price to be paid for electrical energy by the tramways companies to their local Corporations.

Hammond's experience in building and running supply stations made him an authority in rating appeals for electrical works and he was a familiar figure as an 'expert witness' in Parliamentary Committee Rooms. He acted in a professional capacity in nearly all the inquiries relating to the electrical power distribution Bills.

As was to be expected for a man with such pioneering knowledge and experience he wrote many technical articles and papers and after successfully setting up his electrical engineering college he also founded (in 1890) the Electrical Standardising, Testing and Training Institution. Hammond's capabilities were recognised by his contemporaries and he was a member of the Institutes of both the Civil Engineers and the Electrical Engineers and in 1902 he became the honorary treasurer of the IEE. On 5 August 1915, after a short illness, Robert Hammond died, aged sixty-five.

Notes and references

1. *Leeds Mercury,* 2 February 1899.
2. *Electrician*, 11 March 1898.
3. *Ibid*, 11 March 1898.
4. *Ibid*, 25 August 1899.
5. *Ibid*, 26 February 1897.
6. *Ibid*, 9 March 1900.
7. *Ibid*, 18 March 1910.
8. *Ibid*, 6 March 1908.
9. *Ibid*, 18 March 1910.
10. *Ibid*, 17 March 1911.
11. *Ibid*, 22 March 1912.
12. *Ibid*, 20 March 1914.
13. *Ibid*, 6 April 1917.
14. *Ibid*, 22 March 1918.
15. *Ibid*, 6 April 1917.
16. *Ibid*, 15 November 1918.
17. *Ibid*, 3 March 1922.
18. *Ibid*, 3 February 1911. Total load connected nationally 1,492,277 kW. Load connected to stations supplying only AC current 209, 906 kW.
19. *Electrician*, 16 March 1928.
20. *Ibid*, 26 September 1902.
21. *Electrician,* 11 July 1919.
22. *Ibid*, 12 December 1913.
23. According to *The Times* of 15 June 1898 the MP for Ormskirk told Parliament that this country was 'sadly behindhand in the matter of electricity. It was a century behind the United States in this respect.' Electricity was used less in England than elsewhere in the world. In 1908 the amount of electricity used for power purposes per head of population was 9.3 units in London, 20.7 in Berlin and 24.3 in Chicago (*The Organisation of Electricity Supply in Great Britain*, H. H. Ballin, p. 71).

    In 1918 the Parsons Committee reported that the existing system of electrical generation and distribution was behind the times and was therefore a serious handicap in international competition. (Parsons Report of the Departmental Committee appointed by the Board of Trade to consider the position of the electrical trades after the war, Cmd. 9072, 1918; BPP 1918, XIII, 355).
24. In 1913 total connections in Britain were 1,718, 960kW. Of these 1,218, 585 kW were connections to municipal stations – *Electrician*, 7 February 1913.
25. *Electrician,* 4 December 1903.
26. *Ibid*, 31 July 1914, 1 January 1915, 26 February 1915.
27. *Ibid,* 24 June 1921.
28. *Ibid*, 4 March 1921 – £42,903, 15 December 1922 – £41.041, 23 March 1923 – £51,221.
29. *Ibid*, 25 December 1925 – 'A skeleton of the body envisaged in the Act.'
30. *Electrician,* 26 March 1926.
31. *Ibid*, 17 May 1929.

32. *Ibid*, 14 April 1932.
33. *Ibid*, 6 April 1928.
34. *Ibid*, 19 April 1929, 15 February 1929 – 'Interconnecting main 22,000 volts.'
35. *Ibid*, 19 April 1929.
36. *Ibid*, 4 April 1930.
37. *Ibid*, 17 May 1929.
38. *Ibid*, 3 April 1931.
39. *Ibid*, 1 April 1932 – units increase 13%, 7 April 1932 – units increase 16%.
40. *Ibid*, 16 March 1928 – 595 separate undertakers at end of 1925, 14 June 1935 – 376 local authority and 266 private company undertakings.
41. *Ibid*, 8 March 1935.
42. *Ibid*, 6 September 1935.
43. *Ibid*, 29 March 1935.
44. The 'potted' history of Robert Hammond is taken from obituaries published in the *Yorkshire Post* on 9 August 1915; the *Electrician*, 6 August 1915; and *The Times*, 7 August 1915.

Chronology

1878		Gladstone adopted as Leeds Parliamentary Candidate.
1880	April	Gladstone elected to represent both Leeds and Mid-Lothian in Parliament.
	May	Gladstone's son elected at Leeds in his place.
	August	Herbert Gladstone appointed Junior Lord of Treasury, re-election required. Political rally at Alf. Cooke's. Probable first public display of electric light in Leeds.
1881		Parsons working at Kitsons.
	July	Report by Leeds architect on electric light for new Municipal Buildings – advocated delay until buildings nearer completion.
	July–December	Hammond buys exclusive Brush rights for Yorkshire and nine other counties.
	September	Hammonds street lighting at Chesterfield.
	October	Gladstone's visit to Leeds – rally and banquet at converted Coloured Cloth Hall.
		Parson's epicycloidal engine used to drive generator for electric light in Boar Lane.
	December	Hammonds street lighting at Brighton – first continually surviving public supply system in England.
1882	January	Prospectus for Hammond Electric Light and Power Supply Co.
	February	Joint deputation from Leeds to consider electric light for Municipal Buildings.
	May	Joint deputation recommended delay.
		Yorkshire Brush established.
	July	Main fabric of Municipal Buildings finished.
	August	Electric Lighting Act.
		Appointment of Leeds ELC.
	September	First meeting of ELC – asks for tenders of electric light in

		Municipal Buildings. Decides also to recommend Council's application to Board of Trade.
	November	Council meeting confirms Provisional Order application. Asks for Provisional Order to be permissive, not obligatory – refused.
	December	Leeds joint deputation to President of Board of Trade – ammendments rejected.
		Ferranti's machine taken on by Hammonds.
1883		Parsons leaves Kitsons.
	January	ELC considers replies to tender applications for Municipal Buildings installation. Second deputation formed to investigate.
	April	Report of second deputation. Recommends that ELC ask for new tenders.
		ELC discusses Provisional Order – limited support.
		Indecision by Council – to meet next month.
	May	Council rejects Provisional Order; approves lighting of Municipal Buildings. ELC again asks for tenders.
	September	Yorkshire Brush given contract, but ceases trading. ELC decides not to proceed with electric light – uproar – electric light expected for Music Festival.
	October	Yorkshire Brush wound up – amalgamated with Hammonds. Electric light in Town Hall achieved in three weeks by Cromptons.
		Arc lamps added in Victoria Square for Liberal Conference.
	November	ELC re-elected after protests – appoints sub-committee 'to investigate the means of generating power'.
1884		Parsons, at Clarke, Chapman invents turbine.
	January	ELC recommends gas engines – two bought – to complete by 1 May.
	March	Publicly realised that electric light would not be ready for opening of Municipal Buildings.
	April	Crompton's asked to re-commission electric light in Town Hall ready for opening of Municipal Buildings.
		Hammonds increase capital in attempt to survive.
	17 April	Municipal Buildings opened.
	June	Electric light at last ready in Municipal Buildings.
	December	Electric light in Municipal Buildings unsatisfactory – litigation arranged for summer 1885.
1885		Incandescent Gas Mantle – Karl Auer.
	June	Hammonds wound up.
	July	Leeds obtains out-of-court settlement – new dynamos and lamps supplied.
		Prince and Princess of Wales open new buildings of Yorkshire College – lunch in Coliseum, lit by electric light.

	December	New electric light system in Municipal Buildings proves effective.
1886	July	Arc lamps installed in Town Hall.
	November	Re-election of ELC again queried.
1887	March	Tenders accepted for erection of new Fine Art Gallery to be lit by incandescent lamps.
	November	Attempt to stop re-election of ELC.
1888	February	ELC choose compressed air engines for Fine Art Gallery.
	21 February	Advertisement for new company – House-to-House Electric Light Supply Co. Ltd.
	June	Compressed air motors not available – ELC choose Willans high speed steam engines.
	28 June	Electric Lighting Act 1888.
	September	New Reading Room lit by electric light.
	3 October	New Fine Art Gallery opened – electric light ready.
		ELC asks Borough Engineer to prepare scheme to light centre of Leeds.
	November	Objections to re-election of ELC.
1889	February	House-to-House open new station, West Brompton, London.
	May	Borough Engineers report presented – cost £40,000.
	June	ELC accepts Borough Engineers report – also to recommend extension of electric light in Municipal Buildings £3,000.
	26 June	Yorkshire House-to-House registered.
	July	Council considers town lighting. Accepts extension in Municipal Buildings.
	September	Country's first municipal power station, Bradford.
	October	Decision on Leeds town lighting delayed six months.
1890	January	Leeds Council discusses town lighting, before expiration of six months.
	March	Board of Trade tells Leeds Provisional Order applications from private companies could stand a year.
	July	Leeds gas strike – electric light installed in *Yorkshire Post*.
	October	Council considers ELC report on electric light. Proposal to apply for Council Provisional Order rejected – to consider Provisional Orders from private companies.
1891	April	Council agrees to Provisional Order of Yorkshire House-to-House.
	May	Complaints about electric light in Town Hall and Municipal Buildings.
	July	Yorkshire House-to-House Provisional Order confirmed.
	December	Electric light in Municipal Buildings described as 'fitful'.
1892	January	Insurance for electric light in Munipal Buildings refused until refurbished. £500 spending agreed.
	March	Yorkshire House-to-House prospectus issued – shares sold.

	October	Board of Trade confirms technical approval of Yorkshire House-to-House system.
	14 December	Yorkshire House-to-House suffers building delays – installs temporary generator.
1893	January	Sanitary Committee complains of dense smoke from chimneys of Municipal Buildings.
	1 May	Permanent supply starts from Yorkshire House-to-House station, Whitehall Road – 250 kW.
	10 May	Official opening of Yorkshire House-to-House station.
	November	Extension of Yorkshire House-to-House station to 350 kW.
1894	January	Yorkshire House-to-House offers to supply Municipal Buildings.
	September	ELC recommends acceptance of Yorkshire House-to-House offer for Municipal Buildings.
		Extension of Yorkshire House-to-House station to 750 kW.
	October	Council accepts ELC recommendation.
1895		Bill to extend Leeds tramways.
	Spring	Extension of Yorkshire House-to-House station to 900 kW.
	March	Leeds Corporation loses power to issue 'irredeemable stock'.
		Town Hall connected to Yorkshire House-to-House system.
1896	May	Yorkshire House-to-House offers to supply Leeds trams.
	July	Yorkshire House-to-House offer rejected.
	December	Yorkshire House-to-House offers to supply street lighting.
1897	March	Yorkshire House-to-House repeats offer to supply street lighting.
	May	Yorkshire House-to-House offer rejected.
	September	Council debates takeover of Yorkshire House-to-House. Left to Parliamentary Committee.
	October	Extension to Yorkshire House-to-House station to 1500 kW. 900 kW more proposed.
	December	High Court rules 'annuity' means 'in perpetuity'.
1898	February	Leeds Council offers £210 redeemable stock for every £100 capital of Yorkshire House-to-House.
	March	Yorkshire House-to-House rejects Council's offer.
		Local Government Board inquiry in Leeds to consider issue of irredeemable stock.
	May	LGB confirms issue of Provisional Order to allow issue of irredeemable stock.
	July	Provisional Order receives Royal Assent. Council negotiations with Yorkshire House-to-House Agreed: Capital takeover: 30 September, Completion of purchase: 15 November.
	8 November	Agreement between Council and Yorkshire House-to-House.
	9 November	Council approves agreement.
	21 November	Yorkshire House-to-House Extraordinary General Meeting approves agreement.

	15 December	Completion of purchase.
1899	24 January	Yorkshire House-to-House final Annual Meeting and Extraordinary General Meeting – resolution passed to wind up company.
	10 February	Yorkshire House-to-House Extraordinary General Meeting confirms winding-up. Chairman appointed liquidator.
	June	Irredeemable stock sold at £170 per cent.
1900	March	Payments to Yorkshire House-to-House shareholders complete.
1902	September	Leeds Corporation's new generating station opens at Whitehall Road – AC 50 c/s 2 phase. Plant now 8,740 kW (3,000 kW in new station).
1915	5 August	Hammond dies.

Appendix

Electricity supply undertakings: an analysis of the first twenty-five years, 1882 to 1907.

Undertakings are shown as Municipally (M) or Privately (P) owned at the time of commencement of supply. Those undertakings transferred from private to municipal ownership by the end of 1907 are shown PM. Undertakings giving supply to Tramways in 1907 are shown T. A dash (–) indicates that no information is available.

Date of commencement of supply	Location of undertaking	Capital Expended to end 1896[1] (£000s)	At end 1907				
			ownership	AC or DC supply	Tramway supply	Plant capacity[2] (MW)	Number of consumers[3]
1882	Eastbourne	50	PM	A		2.0	1269
	Hastings	49	PM	A		1.9	844
1883	Liverpool	300	PM	D	T	28.3	6607
1885	London – Deptford	900	P	A/D		17.3	2390
1886	Taunton	23	PM	A/D	T	5.1	970
1887	Kensington/Knightsb'ge	220	P	D		2.3	4183
	Leamington	32	P	D		0.5	318
1888	Chatham/Rochester	20	P	A		3.5	426
1889	Bath	45	PM	A/D		1.9	686
	Bradford	109	M	D	T	9.7	2259
	Brompton/Kensington	134	P	A		2.6	4127
	Chelsea	215	P	D		3.5	–
	Exeter	11	M	A/D	T	1.5	1022
	Galway	8	P	D		0.2	106
	Lynton	8	P	A		1.8	116
	Newcastle	76	P	A/D	T	39.5	4810
	Reading	35	P	A/D		2.7	940
	St. James/Pall mall	241	P	D		7.9	2455
1890	Chelmsford	19	P	A		0.9	731
	Fareham	9	M	A		0.1	189
	Keswick	5	P	A		0.1	–
	London – Metropolitan	680	P	A/D		19.2	9624
	Newcastle	67	P	A/D		7.4	1250

	Westminster	482	P	D		8.0	–
	Woking	15	P	A		1.2	1109
1891	Birmingham	150	M	A/D	T	19.5	3500
	Brighton	174	M	A/D	T	9.8	4500
	Bournemouth (Poole)	72	P	A/D	T	4.0	3634
	Charing Cross	310	P	A/D		20.4	–
	City of London	1,202	P	A/D		23.0	–
	Larne	10	P	A		0.2	120
	Northampton	22	P	D		1.1	1075
	Notting Hill	99	P	D		0.6	2579
	Prescot	12	P	A/D	T	1.8	375
	St. Pancras	165	M	A/D		6.0	3111
	Southampton	27	M	A/D	T	2.2	744
1892	Bray	10	M	A		0.3	215
	Cambridge	38	P	A		2.0	861
	Dublin	53	M	A		6.0	3627
	Hove	62	P	D		1.6	1436
	Nelson	10	M	D	T	1.0	350
	Ogmore Valley	5	P	A/D		0.1	320
	Oxford	75	P	A/D		1.6	1475
	Preston	79	P	A/D		2.4	1300
1893	Blackpool	80	M	A/D	T	4.2	1108
	Bristol	110	M	A/D		9.1	3000
	Burnley	24	M	D	T	2.0	657
	Derby	50	M	A/D	T	5.1	1589
	Dundee	35	M	D	T	3.0	1460
	Glasgow	135	M	D		27.1	15497
	Hamilton	–	P	D		0.4	300

Date of commencement of supply	Location of undertaking	Capital Expended to end 1896[1] (£000s)	At end 1907				
			ownership	AC or DC supply	Tramway supply	Plant capacity[2] (MW)	Number of consumers[3]
	Huddersfield	57	M	A		3.2	2630
	Hull	47	M	D		6.8	2864
	Killarney	5	P	A		0.2	110
	Kingston-on-Thames	24	M	A		0.9	824
	Leeds	105	PM	A		13.7	6100
	Manchester	287	M	A	T	34.5	6989
	Norwich	66	PM	D		3.6	4200
	Pontypool	7	P	D		0.2	154
	Richmond (Surrey)	39	P	D		1.1	1426
	Scarborough	33	P	A/D	T	1.9	900
	Sheffield	87	PM	A		7.4	3500
	Whitehaven	19	M	D		0.4	385
	Woolwich	15	M	A/D		2.5	600
1894	Aberdeen	30	M	D	T	4.6	1779
	Altrincham	28	P	A		0.7	1022
	Bedford	44	M	A		1.3	1458
	Bolton	48	M	A/D	T	5.9	2221
	Burton-on-Trent	29	M	A/D	T	1.7	473
	Cardiff	50	M	A/D	T	7.0	1465
	Coatbridge	19	P	A/D	T	1.9	–
	Dewsbury	24	M	D	T	1.0	384
	Ealing	45	M	A		1.8	3271
	Halifax	46	M	A/D	T	4.8	1550

	Hampstead	80	M	A		4.8	5714
	Hanley	44	M	A		1.7	950
	Lancaster	25	M	D	T	1.0	800
	Leicester	38	M	A		4.4	2741
	Londonderry	20	M	D		0.7	260
	Nottingham	62	M	D		7.4	3392
	Oldham	32	M	D	T	5.2	858
	Portsmouth	87	M	A		3.3	3950
	Southport	46	M	A/D	T	2.4	1780
	Worcester	60	M	A/D	T	1.9	780
	Yarmouth	28	M	A/D	T	1.7	1300
1895	Aberystwyth	–	P	D		0.4	361
	Belfast	35	M	A/D	T	6.8	1800
	Blackburn	35	M	A/D	T	4.5	1670
	Cheltenham	26	M	A/D	T	2.0	944
	Coventry	24	M	A		3.3	994
	Dover	48	PM	A/D	T	1.9	829
	Edinburgh	192	M	A/D		12.8	10432
	Newport (Mon)	37	M	A/D	T	3.3	1121
	Oswestry	5	P	D		0.2	198
	Salford	43	M	A/D	T	7.5	1362
	Shrewsbury	17	PM	D		0.8	520
	Stafford	17	M	D		0.6	262
	Tunbridge Wells	25	M	A		1.3	830
	Walsall	19	M	D	T	1.7	522
	Wolverhampton	35	M	A/D	T	5.3	1081
1896	Ayr	27	M	A/D	T	1.3	1098
	Birkenhead	16	M	D		1.4	1108

Date of commencement of supply	Location of undertaking	Capital Expended to end 1896[1] (£000s)	At end 1907				
			ownership	AC or DC supply	Tramway supply	Plant capacity[2] (MW)	Number of consumers[3]
	Bury	–	M	D	T	1.8	488
	Chester	25	M	D	T	1.3	1020
	Croydon	28	M	A/D	T	5.7	2000
	Guildford	6	P	D		0.5	600
	Harrow	–	P	D		0.5	1050
	Islington	102	M	A		5.1	1660
	Leyton	18	M	D		2.6	1673
	St. Helens	9	M	D	T	1.8	530
	South Shields	29	M	A/D	T	2.7	1144
	Sunderland	38	M	A/D	T	6.8	1170
	Windsor & Eton	–	P	A/D		0.6	665
1897	County of London (north)	–	P	A/D		5.6	–
	Harrogate	–	M	A		1.9	1243
	Hammersmith	38	M	A		5.0	2098
	Liverpool district	–	P	D		0.6	578
	Northwich	–	P	D		0.2	452
	Nuneaton	–	M	D		0.6	326
	Shoreditch	30	M	D		3.0	920
	Smithfield markets	–	P	D		1.0	390
	Wakefield	–	M	A		1.6	580
	Wallasey	18	M	A/D	T	2.1	1420
1898	Bootle		M	D	T	2.1	630
	Colchester		M	D	T	2.0	1263

	Cork		P	A/D	T	1.7	1550
	County of London (south)		P	A		5.5	–
	Eccles		M	A		0.5	238
	Folkestone		P	D		1.3	2335
	Gillingham		M	A/D		0.8	306
	Lincoln		M	D	T	1.4	525
	Llandudno		M	D		1.0	459
	Morecambe	6	M	D		0.7	322
	Morley		M	A		0.2	275
	Salisbury		P	D		0.6	1013
	Stockport		M	D	T	2.4	–
	Torquay	22	M	A/D	T	1.1	427
	West Ham		M	A/D	T	8.4	1950
	Winchester		P	D		0.9	1010
	Wycombe		P	D		0.8	1540
1899	Alderley Edge		P	D		0.2	350
	Ashton-u-Lyne		M	D	T	1.8	405
	Barking		M	A/D	T	1.0	800
	Barrow		M	D	T	1.5	712
	Bromley		P	D		1.6	1300
	Canterbury		M	D		0.9	540
	Carlisle		M	D	T	0.9	440
	Darwen		M	D	T	0.9	371
	Dudley		M	D	T	1.6	336
	Felixtowe		M	D		0.2	365
	Greenock		M	A/D	T	3.5	692
	Hereford		M	D		0.8	244
	Kings Lynn		M	D		0.7	613

Date of commencement of supply	Location of undertaking	Capital Expended to end 1896[1] (£000s)	At end 1907				
			ownership	AC or DC supply	Tramway supply	Plant capacity[2] (MW)	Number of consumers[3]
	Lambeth		P	A		3.8	2200
	Leith		M	D	T	2.8	1301
	Monmouth	6	M	A		0.1	130
	Newmarket		P	D		0.4	283
	Newton Abbott		P	D		0.2	275
	Northallerton		P	D		0.1	123
	Paisley		M	A/D	T	3.8	722
	Plymouth		M	A/D	T	2.8	880
	Redditch		M	A		0.5	530
	Rothesay		P	D		0.1	–
	Saltburn		P	D		0.1	150
	Southwark		M	D		1.2	560
	Stepney		M	D		3.7	940
	Ventnor		P	D		0.3	590
	Watford		M	A		1.1	588
	Wigan		M	D	T	2.9	600
	Wimbledon		M	A		2.7	2945
1900	Accrington		M	D	T	0.9	383
	Bangor		M	D		0.4	430
	Barnsley		M	D	T	1.4	720
	Beckenham		M	A		1.4	1200
	Bexhill		M	D		1.2	680
	Bournemouth		M	D	T	1.3	–

Bury St. Edmunds	M	D		0.2	250
Buxton	M	D		0.6	300
Chiswick	P	D		0.9	1220
Crewe	M	D		0.7	460
Darlington	M	D	T	1.2	606
Doncaster	M	D	T	1.2	490
Fleetwood	P	D		0.4	225
Gloucester	M	D	T	1.9	535
Govan	M	D		2.0	948
Guernsey	P	D		0.9	—
Hindhead	P	D		0.2	120
Ingleton	P	D		0.1	85
Kidderminster	P	D	T	1.4	407
Leigh	M	D		0.8	315
Lymington	P	D		0.1	398
Manchester (Trafford Park)	P	A/D	T	3.1	784
Melton Mewbray	P	D		0.3	538
Middlesborough	M	D		1.6	790
Notting Hill	P	A		6.0	—
Peterborough	M	D	T	0.8	325
Poplar	M	A/D		3.4	474
Queenstown	P	D		0.3	203
Rathmines	M	D		0.8	858
Rochdale	M	A/D	T	2.0	380
South Metropolitan	P	A		3.0	3201
Stirling	M	D		0.6	380
Stockton-on-Tees	M	D		1.1	310
Warrington	M	A/D	T	2.2	320
Witney	P	D		0.2	100
York	M	D		1.8	660

Date of commencement of supply	Location of undertaking	Capital Expended to end 1896[1] (£000s)	At end 1907				
			ownership	AC or DC supply	Tramway supply	Plant capacity[2] (MW)	Number of consumers[3]
1901	Barnes		M	D		1.0	1300
	Battersea		M	D		3.6	1066
	Blyth		P	A/D		0.9	1070
	Brechin		P	D		0.3	460
	Chesterfield		M	D	T	1.0	524
	Colwyn Bay		M	D		0.4	320
	Colne		M	D	T	1.1	290
	Durham		P	A/D		0.4	262
	East Ham		M	D	T	1.9	944
	Farnworth		M	D	T	0.6	306
	Fulham		M	A		3.0	2177
	Grays Thurrock		M	D		0.4	279
	Grimsby		M	D	T	1.4	522
	Hackney		M	D		3.3	1360
	Heckmondwike		M	D	T	1.0	239
	Hertford		P	D		0.3	265
	Heywood		M	D	T	0.6	168
	Hoylake		M	A		0.6	910
	Ilford		M	D	T	2.6	3000
	Keighley		M	D	T	0.7	274
	Lewes		P	D		0.2	180
	Longton		M	D		0.3	111
	Lowestoft		M	D	T	1.6	797

	Luton	M	D		0.6	360
	Maidstone	M	D	T	0.8	658
	Merthyr Tydfil	P	D	T	0.8	396
	Montrose	P	D		0.3	495
	Motherwell	M	D		1.9	534
	Newcastle	M	D	T	5.2	—
	Newport/Cowes	P	D		0.4	835
	Penarth	P	D		0.6	290
	Perth	M	D	T	1.2	357
	Reigate	M	A		0.5	436
	Rhyl	M	D		0.5	250
	Rotherham	M	D	T	1.2	330
	St. Annes	M	D	T	0.8	640
	Shanklin/Sandown	P	D		0.4	560
	Shipley	M	D	T	1.1	400
	Sleaford	M	D		0.2	125
	Sutton Coldfield	M	D		0.6	485
	Swansea	M	A/D		2.3	1080
	Tynemouth	M	D	T	1.1	706
	Walthamstow	M	D	T	1.6	882
	West Bromwich	M	D	T	1.6	300
	West Hartlepool	M	D		1.6	636
	Weston-Super-Mare	P	D	T	1.1	300
	Worksop	M	D		0.5	658
	Worthing	M	D		0.6	542
	Wrexham	M	D	T	0.7	280
1902	Aldershot	M	D		0.4	178
	Alnwick	P	D		0.2	550

Date of commencement of supply	Location of undertaking	Capital Expended to end 1896[1] (£000s)	At end 1907				
			ownership	AC or DC supply	Tramway supply	Plant capacity[2] (MW)	Number of consumers[3]
	Bermondsey		M	D		1.5	340
	Birkdale		P	D	T	0.4	415
	Broughty Ferry		M	D		0.2	167
	Cambourne/Redruth		P	D	T	1.0	960
	Cleckheaton		M	D	T	0.8	105
	Dartmouth		P	D		0.3	532
	Devonport		M	D	T	1.8	326
	Epsom		M	D		0.3	195
	Godalming		P	D		0.4	704
	Gravesend		M	D	T	1.1	380
	Hawick		P	D		0.6	402
	Horsham		M	D		0.4	420
	Jarrow		P	A/D		0.5	85
	Kendal		M	D		0.1	110
	Kircaldy		M	D	T	1.1	341
	Leatherhead		P	D		0.2	218
	Llandilo		M	D		0.1	100
	Maidenhead		M	D		0.4	430
	Margate/Broadstairs		P	A/D	T	0.5	276
	Market Drayton		P	D		0.1	270
	Mexborough		M	D		0.3	190
	Middleton		M	D	T	0.8	170
	Patrick		M	D		1.5	1260

	Ross	P	D		0.1	79
	Southend-on-sea	M	D	T	1.8	1356
	Spennymoor	P	D		0.4	1277
	Stamford	P	D		0.4	634
	Sutton/Carshalton	P	A/D	T	1.6	595
	Tadcaster	P	D		0.1	–
	Tonbridge	M	D		0.1	190
	Twickenham	P	D		0.8	3268
	Uxbridge	P	A		1.0	395
	Weybridge/Walton	P	D		0.6	1200
	Whitby	M	D		0.4	360
1903	Aston Manor	M	A/D	T	2.5	500
	Banbury	P	D		0.4	206
	Barnstaple	M	D		0.4	302
	Batley	M	D		0.9	170
	Berwick-on-Tweed	P	D		0.2	290
	Bexley	M	A/D	T	0.8	327
	Caterham	P	D		0.2	270
	Christchurch	P	D		0.4	223
	Cromer	P	D	T	0.2	338
	Elland	M	D		0.3	210
	Erith	M	A	T	1.4	696
	Falkirk	M	D		0.5	308
	Finchley	M	D		1.0	1396
	Frinton-on-sea	P	D		0.1	175
	Glossop	P	D	T	0.4	350
	Grantham	P	D		0.4	1164
	Hebburn	P	A/D		1.9	849

Date of commencement of supply	Location of undertaking	Capital Expended to end 1896[1] (£000s)	At end 1907				
			ownership	AC or DC supply	Tramway supply	Plant capacity[2] (MW)	Number of consumers[3]
	Honley		M	D		0.1	60
	Hornsey		M	D		1.3	1259
	Ilfracombe		P	D		0.3	222
	Ilkeston		M	D	T	0.7	200
	Ipswich		M	D	T	1.5	372
	Kilmalcolm		P	D		0.1	160
	Limerick		M	D		0.2	225
	Long Easton		M	D		0.5	295
	Mansfield		M	D		0.7	400
	Milford-on-sea		P	D		0.1	70
	Oban		M	D		0.2	130
	Roundhay (Leeds)		P	D		0.2	200
	Ryde (I of W)		P	A/D		0.2	595
	Sheerness		P	A/D	T	0.5	187
	Slough/Datchet		P	D		0.6	360
	Stretford		M	D	T	1.3	202
	Surbiton		M	D		0.4	390
	Swindon		M	D	T	0.9	630
	Whitchurch		P	D		0.1	66
	Wormit (Tayside)		P	D		0.1	70
1904	Bridgewater		P	D		0.2	140
	Chepstow		P	D		0.1	53

Chesham	P	D		0.3	191
Church Stretton	P	D		0.1	61
Dalkeith N.B.	P	D		0.1	83
Dollar N.B.	P	D		0.1	90
Dorking	P	D		0.2	400
Faversham	M	D		0.3	145
Frome	P	D		0.5	359
Hartlepool	P	D		0.2	47
Hebden Bridge	M	D		0.2	120
Heston/Isleworth	M	D		0.6	513
Holyhead	M	D		0.2	292
Jedburgh N.B.	P	D		0.1	85
Kettering	M	D		0.5	432
Kildare	M	D		0.1	46
Kilmarnock	M	D		0.8	171
Leek	M	D		0.3	145
Llangollon	P	D		0.1	41
Loughborough	M	D		0.5	236
Malton/Norton	P	D		0.1	137
Malvern	M	A		0.2	156
Melrose N.B.	P	D		0.1	44
Morpeth	P	D		0.1	108
Musselburgh	P	D	T	0.6	323
Newcastle-u-Lyme	M	D		0.1	64
Radcliffe	M	D	T	0.5	61
Smethwick	P	D	T	3.5	250
Stalybridge	M	A/D	T	5.5	260
Stoke-on-Trent	M	D		0.6	306

Date of commencement of supply	Location of undertaking	Capital Expended to end 1896[1] (£000s)	At end 1907				
			ownership	AC or DC supply	Tramway supply	Plant capacity[2] (MW)	Number of consumers[3]
	Thirsk/Sowerby		P	D		0.1	74
	Totnes		P	D		0.1	97
	Weymouth		M	D		0.3	309
1905	Bowness		P	D		0.3	330
	Bridlington		M	D		0.3	280
	Burslem		M	D		0.2	200
	Cambuslang		M	D		0.1	120
	Caernarvon		P	D		0.2	210
	Exmouth		P	D		0.2	168
	Handsworth		M	D		0.2	182
	Inverness		P	D		0.2	199
	Newbury		P	D		0.4	402
	Pontypridd		M	D	T	0.9	283
	Ramsgate		P	D		0.3	306
	St. Andrews N.B.		P	D		0.3	161
	St. Marylebone		M	D		12.0	6600
	Shildon		P	D		0.1	277
	Todmorden		M	D		0.2	90
	Warwick		P	D	T	0.6	111
1906	Clacton-on-sea		M	D		0.2	47
	Dumbarton		P	D	T	0.4	142
	Dumfries		P	D		0.2	58

	Falmouth	P	D		0.3	69
	Hexham	P	A		0.1	60
	Ludlow	P	D		0.1	–
	Newquay	P	D		0.2	127
1907	Ascot	P	D		0.1	20
	Dartford	M	D	T	0.8	410
	Pembroke (Ireland)	M	D		0.3	430
	Stratford-on-Avon	P	D		0.1	–
	Wellingborough	P	D		0.3	220

Notes:
[1] From Supplement to *Electrician*, 26 February 1897 (to nearest £1,000).
[2] From Supplement to *Electrician*, 24 January 1908 (to nearest 0.1 MW).
[3] From Supplement to *Electrician*, 24 January 1908.

Other undertakings

At the end of 1907 there were also fourteen power companies, supplying mostly other undertakings which did not generate their own electricity. There were twenty-eight such undertakings being supplied 'in Bulk'. It is also possible that undertakings existed in 1893 and 1894 in the following places:

Buckingham
Carlow
Chagford
Cowpen
Crystal Palace
Kelvinside
Keynsham
Morecambe
St. Austell
St. Lukes (London)
Thetford
Waterford
Weybridge

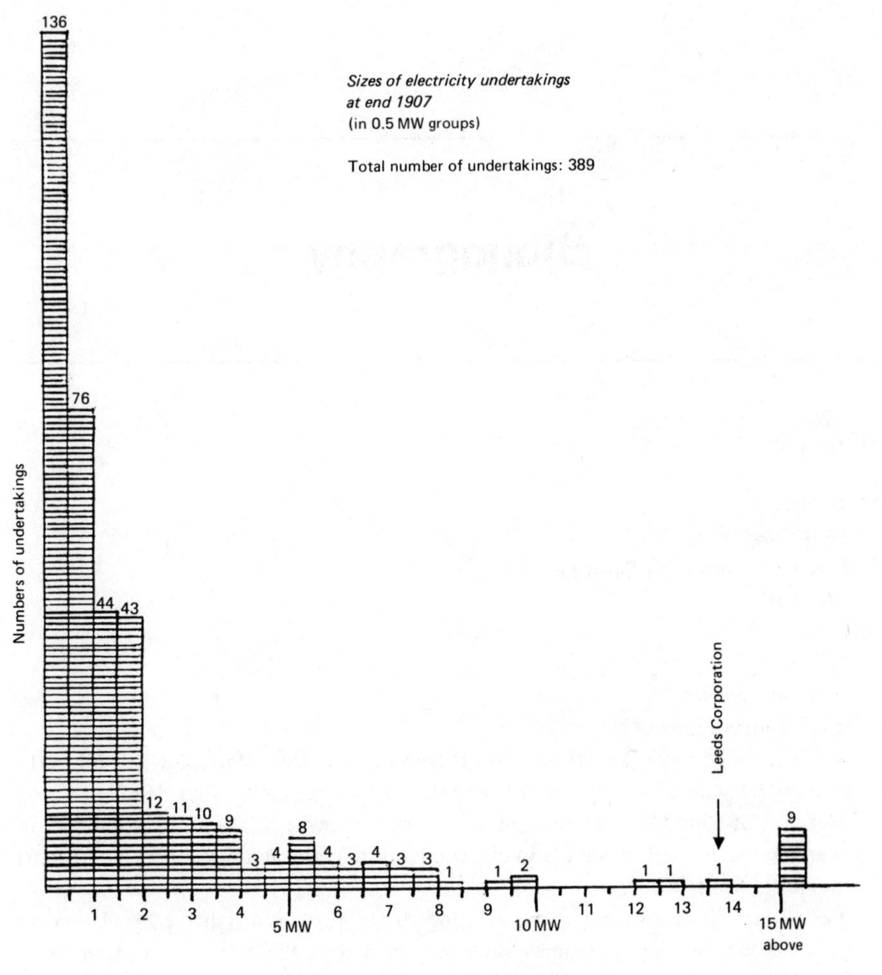

Sizes of electricity undertakings
at end 1907
(in 0.5 MW groups)
Total number of undertakings: 389
Numbers of undertakings
136
76
44
43
12
11
10
9
3
4
8
4
3
4
3
3
1
1
2
1
1
1
9
Leeds Corporation
1
2
3
4
5 MW
6
7
8
9
10 MW
11
12
13
14
15 MW
above
Plant capacity

Bibliography

Major sources

Newspapers:
Yorkshire Post
Leeds Mercury (and Saturday Supplement)
The Times
Magazines:
Engineer
Electrical Review
Leeds Corporation Reports:
'Electric Lighting: Report of Joint Deputation from Purchase of Property Committee, Gas Committee and Free Library Committee' (1 May 1882).
'Report of Sub-Committee appointed by Electric Lighting Committee to examine recent improvements in electric lighting plant and machinery and report thereon' (April 1883).
'Particulars to accompany plans relating to the proposed lighting by electricity of portions of the Municipal Buildings at Leeds for the Corporation, with Supplemental Particulars' (1883).
Parliamentary Report:
'Select Committee on Lighting by Electricity' (1879).

Other sources

Magazines:
Brown, E. J., 'An Outline History of the University of Leeds', Society of Chemical Industry Annual Meeting, Leeds (brochure), (1957).
Laithwaite, E. R., 'The Greatest Experimenter', *Electrical Review,* Volume 25, Number 18, 9 November 1979.
North Eastern Gen. (house journal of the North-East region of the Central Electricity Generating Board), June 1968.

Phillips, V. J., 'Physiological Radio Receivers', *IEE Electronics and Power*, May 1981.

Simpson, R. J. and Power, H. M., 'Electric Arc Light – an episode in the history of feedback and control', *IEE Electronics and Power,* September 1979.

Reference:

The British Encyclopaedia, Odhams Press Ltd., London, 1933.

Dictionary of National Biography, Oxford University Press.

Electricity Supply in Great Britain (a chronology), Electricity Council, April 1973.

History:

Ballin, H. H., *The Organisation of Electricity Supply in Great Britain*, Electrical Press Ltd., London, 1946.

Bowers, B., *A History of Electric Light and Power*, Peter Peregrinus Ltd., UK and New York, 1982.

Clark, Edwin Kitson, *Kitsons of Leeds 1837–1937*, The Locomotive Publishing Co. Ltd., London.

Clark, Edwin Kitson, 'Sir Charles Parsons', *The Post Victorians*, Ivor Nicholson and Watson, Ltd., London, 1933.

Hirst, Stuart (Ed), *The Leeds Tercentenary Official Handbook*, Tercentenary Executive Leeds, 1926.

Parsons, R. H., *The Early Days of the Power Station Industry*, Cambridge University Press, 1940.

Passer, Harold C., *The Electricity Manufacturers 1875–1900*, Harvard University Press, 1953.

Scott, J. D., *Siemens Brothers 1858–1958*, Weidenfeld and Nicolson, 1958.

Self, Sir Henry and Watson, Elizabeth M., *Electricity Supply in Great Britain–its development and organisation*, George Allen and Unwin Ltd., (National Board Series), 1952.

Swan, M. E. and K. R., *Sir Joseph Swan*, Oriel Press, Newcastle, 1968.

Thompson, S. P., *Michael Faraday, his life and Work*, Cassell and Co. Ltd., London (The Century Science Series), 1898.

Private sources

(At Yorkshire Electricity Board, West Yorkshire Area Offices)

Scrapbook of newspaper and magazine cuttings compiled from 1893 to 1898 by Yorkshire House-to-House Co; from 1898 onwards by Leeds Corporation Electricity Department.

Records of mains and details of service connections, Yorkshire House-to-House Co.

Copy of Yorkshire House-to-House Prospectus, dated 26 March 1892, (including Memorandum of Association of Yorkshire House-to-House Electric Co. Ltd.)

Provisional Order for Leeds Electric Lighting by Yorkshire House-to-House Electric Co. Ltd., dated 1891.

Copy of 'Regulations and Conditions for securing the safety of the public and for ensuring a proper and sufficient supply of Electrical Energy, made by the Board of Trade under the provisions of the Electric Lighting Acts, 1882 and 1888, and of the Leeds Electric Supply Order, 1891' dated 18 October 1892 (with attachment: 'Description of the system to be adopted by the Yorkshire House-to-House Electric Co. Ltd., for the supply of energy under the Leeds Electric Supply Order, 1891').

Index

WITHDRAWN
621.32
P87
Mechanics' Institute Library
621.32 P87
dewey
Poulter, J. D./An early history of elect
3 1750 00087 4993
Mechanics' Institute
1855
MECHANICS' INSTITUTE LIBRARY
57 POST ST., SAN FRANCISCO, CALIFORNIA 94104
PHONE 421-1750